The IMA Volumes in Mathematics and its Applications

Volume 103

Series Editors
Avner Friedman   Robert Gulliver

Springer Science+Business Media, LLC

Institute for Mathematics and its Applications IMA

The **Institute for Mathematics and its Applications** was established by a grant from the National Science Foundation to the University of Minnesota in 1982. The IMA seeks to encourage the development and study of fresh mathematical concepts and questions of concern to the other sciences by bringing together mathematicians and scientists from diverse fields in an atmosphere that will stimulate discussion and collaboration.

The IMA Volumes are intended to involve the broader scientific community in this process.

Avner Friedman, Director

Robert Gulliver, Associate Director

* * * * * * * * * *

IMA ANNUAL PROGRAMS

1982–1983	Statistical and Continuum Approaches to Phase Transition
1983–1984	Mathematical Models for the Economics of Decentralized Resource Allocation
1984–1985	Continuum Physics and Partial Differential Equations
1985–1986	Stochastic Differential Equations and Their Applications
1986–1987	Scientific Computation
1987–1988	Applied Combinatorics
1988–1989	Nonlinear Waves
1989–1990	Dynamical Systems and Their Applications
1990–1991	Phase Transitions and Free Boundaries
1991–1992	Applied Linear Algebra
1992–1993	Control Theory and its Applications
1993–1994	Emerging Applications of Probability
1994–1995	Waves and Scattering
1995–1996	Mathematical Methods in Material Science
1996–1997	Mathematics of High Performance Computing
1997–1998	Emerging Applications of Dynamical Systems
1998–1999	Mathematics in Biology
1999–2000	Reactive Flows and Transport Phenomena

Continued at the back

Stuart G. Whittington De Witt Sumners
Timothy Lodge
Editors

Topology and Geometry in Polymer Science

With 67 Illustrations

Springer

Stuart G. Whittington
Department of Chemistry
University of Toronto
Toronto, ON M5S 1A1, CANADA

De Witt Sumners
Department of Mathematics
Florida State University
Tallahassee, FL 32306, USA

Timothy Lodge
Department of Chemistry
University of Minnesota
Minneapolis, MN 55455, USA

Series Editors:
Avner Friedman
Robert Gulliver
Institute for Mathematics and its Applications
University of Minnesota
Minneapolis, MN 55455, USA

Mathematics Subject Classifications (1991): 05C10, 05C80, 52A22, 53A04, 53A05, 57M25, 60D05, 60J15, 82B27, 82B41, 82B80, 82D60, 92E10, 92-08

Library of Congress Cataloging-in-Publication Data
Topology and geometry volume / [edited by] Stuart G. Whittington, De Witt Sumners, Timothy Lodge.
p. cm. — (The IMA volumes in mathematics and its applications ; 103)
Papers from a workshop on the topology and geometry of polymers held at the IMA in June 1996.
Includes bibliographical references and index.
ISBN 978-0-387-98580-0 ISBN 978-1-4612-1712-1 (eBook)
DOI 10.1007/978-1-4612-1712-1
1. Polymers—Mathematical models—Congresses. 2. Topology—Congresses. 3. Geometry—Congresses. I. Whittington, Stuart G. II. Sumners, De Witt L. III. Lodge, Timothy. IV. Series: IMA volumes in mathematics and its applications ; v. 103.
QD381.9.M3T66 1998
547'.7'01514—dc21 98-24449

Printed on acid-free paper.

Production managed by Karina Mikhli; manufacturing supervised by Jacqui Ashri.
Camera-ready copy prepared by the IMA.

9 8 7 6 5 4 3 2 1

ISBN 978-0-387-98580-0

FOREWORD

This IMA Volume in Mathematics and its Applications

TOPOLOGY AND GEOMETRY IN POLYMER SCIENCE

is based on the proceedings of a very successful one-week workshop with the same title. This workshop was an integral part of the 1995–1996 IMA program on "Mathematical Methods in Materials Science." We would like to thank Stuart G. Whittington, De Witt Sumners, and Timothy Lodge for their excellent work as organizers of the meeting and for editing the proceedings.

We also take this opportunity to thank the National Science Foundation (NSF), the Army Research Office (ARO) and the Office of Naval Research (ONR), whose financial support made the workshop possible.

Avner Friedman

Robert Gulliver

PREFACE

This book is the product of a workshop on Topology and Geometry of Polymers, held at the IMA in June 1996. The workshop brought together topologists, combinatorialists, theoretical physicists and polymer scientists, who share an interest in characterizing and predicting the microscopic entanglement properties of polymers, and their effect on macroscopic physical properties.

Linear polymer molecules in dilute solution are highly flexible and self-entangled. In more concentrated solutions, or in the melt, there can be important entanglement effects both within and between polymers, and these entanglements can influence the rheological properties of the system as well as the crystallization properties, and hence the properties of the polymeric system in more ordered states. Although polymer scientists have been aware of these problems for more than forty years, it is only recently that the powerful methods of algebraic topology have been used systematically to characterize and describe these entanglements. Starting from the simplest possible system (a ring polymer in dilute solution) one can ask how badly knotted the polymer will be, as a function of the degree of polymerization, the stiffness, the solvent quality, etc. To some extent these questions have been answered by a combination of rigorous mathematical arguments (combining ideas from combinatorics and from algebraic topology) and numerical methods such as Monte Carlo techniques. For instance, the paper in this book by Orlandini *et al* addresses the question of the relative probability of different knot types in a ring polymer. One can also ask about the differential geometry of the polymer and the paper by Bickis investigates this question for a random walk model. As the concentration increases, linking between rings becomes possible and these links will influence the static and dynamic properties of the solution. Several problems associated with linking are treated in this book by Diao and Janse van Rensburg and by Uberti *et al.*

As we pass from dilute solutions to melts the characterization of the entanglements becomes more difficult. The paper by Edwards addresses these problems. Having characterized the entanglement complexity one then needs to know how it will affect rheological properties. For instance, what is the contribution of entanglements to the elastic properties of a rubbery polymer? How do the dynamics of polymers, in solution or in the melt, depend on entanglement?

The conformations of polymers are strongly influenced by any applied geometrical constraints. Polymers behave quite differently in pores or when confined in a slab geometry and their properties in these environments influence their behaviour as, for instance, stabilizers of colloidal dispersions. Approximate theories of colloidal stability have been available for many years, but it is only recently that simple models of polymers in confined geometries have been analysed rigorously. There can be interesting interactions between topological properties and these geometrical constraints. E.g. how does the knot probability in a ring polymer change when the polymer is confined to a pore or slab? These questions are treated in the papers by Soteros and by Tesi *et al.*

Topological problems also occur in the modelling of polymeric membranes. These are closely related to self-avoiding random surfaces, an area in which rapid progress has recently been made, although many important questions still remain open. Closely related are the properties of vesicles where the topology of the surface can have an important influence on the behaviour of the vesicle. The papers by Einstein and Stella, and by Beichl and Sullivan, discuss numerical approaches to these problems, while Janse van Rensburg presents some corresponding rigorous results.

An area where ideas from physics have had a direct influence on knot theory is the topic of *knot energies*. A knot is a simple closed curve in R^3 (which could be a smooth curve or a finite polygon). One associates an energy functional with the curve so that each embedding has an associated energy. The minimum energy over all embeddings with a fixed knot type is a knot invariant. The book contains four articles on this subject. The paper by Kauffman *et al* describes one such functional and the search for its energy minima. Of course, there is tremendous flexibility about the choice of the energy functional and the paper by Ernst *et al* considers desirable properties of the knot energy and compares different choices for it. The papers by Simon, and by Kusner and Sullivan discuss aspects of a closely related issue, the thickness of knots. This is an area which is attracting considerable interest at the present time, with recently observed correspondences between the electrophoretic mobility of knotted DNA circles and the energy of "ideal" knot representatives.

The book gives an account of recent progress in our understanding of polymer geometry and topology, and the physical ramifications of polymer entanglement. We trust that it will be of interest to specialists in this area. We hope that it will also serve to introduce polymer scientists to topological approaches, and topologists to interesting applications of their subject in polymer science.

Stuart G. Whittington

De Witt Sumners

Timothy Lodge

CONTENTS

Foreword v

Preface vii

1. Entanglement Complexity of Polymers

Entanglements of polymers 1
S.F. Edwards

Entropic exponents of knotted lattice polygons 9
Enzo Orlandini, E.J. Janse van Rensburg,
Maria Carla Tesi, and S.G. Whittington

The torsion of three-dimensional random walk 23
Mikelis G. Bickis

2. Knot Energies

Self-repelling knots and local energy minima 29
Louis H. Kauffman, Milana Huang, and Robert P. Greszczuk

Properties of knot energies 37
Y. Diao, C. Ernst, and E.J. Janse van Rensburg

Energy and thickness of knots 49
Jonathan Simon

On distortion and thickness of knots 67
Robert B. Kusner and John M. Sullivan

3. Random Linking

Percolation of linked circles 79
Y. Diao and E.J. Janse van Rensburg

Minimal links in the cubic lattice 89
R. Uberti, E.J. Janse van Rensburg, E. Orlandini,
M.C. Tesi, and S.G. Whittington

4. Effect of Geometrical Constraints

Knots in graphs in subsets of Z^3 101
C.E. Soteros

Topological entanglement complexity of polymer chains
in confined geometries 135
Maria Carla Tesi, E.J. Janse van Rensburg,
Enzo Orlandini and Stuart G. Whittington

5. Surfaces and Vesicles

Survey of self-avoiding random surfaces on cubic lattices:
Issues, controversies, and results 159
T.L. Einstein and A.L. Stella

Computational methods in random surface simulation 175
Isabel Beichl and Francis Sullivan

A model of lattice vesicles.. 193
E.J. Janse van Rensburg

ENTANGLEMENTS OF POLYMERS

S.F. EDWARDS*

Abstract. The interaction of polymers must not allow them to pass through each other. This poses a topological problem which strictly speaking is not solved. Nevertheless much progress can be made using the Gaussian invariant, and the way to do this is reviewed below. It turns out to give a renormalisable quantum field theory, and is effective in understanding experimental results.

1. The mathematics of polymers. Polymer molecules are totally understood in their specification, but their long chain nature introduces highly complex dynamics, for the chains not only have the dynamics of connection but also cannot pass through one another. Much progress has been made on dense systems, but the topological integrity of the molecules is ignored in dilute solutions, both in self or mutual interactions for no better reason than it is difficult to handle.

Suppose we study the configuration of a chain of N segments of length l, freely hinged, and of no thickness with no interactions. This problem is well known and of course completely soluble. For N large i.e. $N^2 >> N$ the distribution of one end with the other end at the origin tends to

$$(1.1) \qquad (\frac{3}{2\pi N l^2})^{\frac{3}{2}} \exp(-\frac{3R^2}{2Nl^2})$$

which is the solution of

$$(1.2) \qquad (\frac{\partial}{\partial s} - \frac{1}{6}\nabla^2)P(R,s) = \delta(R)\delta(s)$$

were $s = Nl$.

To get P, this Fick equation is fine, but it cannot solve the problem of a polymer threading its way through a hole in a plane barrier, for the condition that at some s the polymer is at the hole, and at all other points of the plane the probability of tending the polymer is zero (for it cannot cross the plane except at the hole), and these two conditions are incompatible when using (1.2).

A real polymer of freely hinged segments is difficult to handle mathematically for its mathematical structure is of a continuous function of its arc length $\underline{R}(s)$, which is not differentiable at $s = ml$, m integral, where the tangent R' changes direction abruptly. A model much studied is to allow the locus $R(s)$ to be differentiable, but have its second derivative R'' to change abruptly.

In terms of Wiener integrals, the Fick equation model is to use a prob-

* Polymers and Colloids, Cavendish Laboratory, Cambridge CB3 OHE, UK.

ability for the whole chain of

$$\omega \exp(-\frac{3}{2l\delta/2} \int R'^2(s)ds)$$

and for the 'wom like chain

$$\omega \exp(-\frac{3}{2l\delta/2} \int R''^2 ds) \prod \delta(R'^2 - 1)$$

or more crudely

$$\omega \exp(-\kappa \int R''^2 ds - \gamma \int R'^2 ds)$$

where k represents the energy of bending over kT and γ fits the long range size Nl^2. Such molecules do exist and are important e.g. DNA, but as a model it obviously is a crude representation of say polystyrene.

These matters are dealt with in detail in textbooks(1,2,3) and I assume the reader knows this or will study the books cited.

In order to give the reader an intuitive view of what worries polymer theorists, I pose the problem which will be calculated in section.

Imagine a box full of long random flight loci which are all frozen except one. Hold the ends of that fixed, but otherwise allow it to move in Brownian motion. The surrounding loci will hold our free locus in what is roughly a tube. What is the radius, a of this tube. It can only depend on c the concentration of the other loci, and l, but is not determined by dimensional analysis

$$a = a\left(\frac{L}{V}, 1\right) = a(c, l)$$

L total length of all the loci, V volume of the box:

$$a = l(cl^2)^\alpha$$

What is α?

2. Topology and polymers. For an unambiguous situation consider two perfect circular rings, $R_1(s)$, $R_2(s)$. They are either joined, or free. The criterion given by Gauss is that

$$I = \oint\oint dR_1 \times dR_2 \cdot \nabla |R_1 - R_2|^{-1} \tag{2.1}$$

is zero for free rings, non zero for non free rings. The sign of I depends on the convention assigned to the direction of the tangent. This is already a problem as the molecule does not have a sense for its tangent, and this turns out to be tiresome in applications; but this paper will not get that far.

When one considers more than two rings e.g. the three Borromean rings one finds it easy to get a configuration such that $I_{12} = I_{23} = I_{31} = 0$, but the rings are entangled. Also there is no analogue of the Gaussian integral (2.1) for self knots e.g. the simplest trefoil knot. By projecting the locus of the curves in space onto a surface, and labelling each intersection on that surface according to which curve is nearest to the surface, one can construct Alexander polynomials, and new Vaughan Jones polynomials have recently appeared. These do not seem to be suitable for use in statistical physics however, though there are papers using them in simulations. The power of the gauss integral can be expressed this way:

Suppose we have a box of rings. The $I_{ij} = 0$ for all pairs. Thus the entropy of the system is

$$S = k_B \log \int \exp\left(-\frac{3}{2l}\sum_i \int R_i'^2 ds_i\right) \prod_{\alpha\beta} \delta(I_{\alpha\beta}) \prod_i DR_i. \tag{2.2}$$

This is complete and explicit. One could feed the $R_i(s)$ into a computer and evaluate. More to the point, one has an explicit formula to start from and calculate. Of course this is a peculiar system and would permit higher connections such as the Borromeans, but one can imagine a system which really is represented by (2.2).

One can make rings, but very long polymers will have their topology dominated by local effects. It seems reasonable, particularly if a system is cross linked chemically thus freezing in its topology, that there is a good problem to be studies via (2.2) i.e.

Fig (1) cannot turn into figure (2). The shaded regions represent the rest of the system. This situation is covered by making I_{12} a constant.

To do this requires two further steps. The first is to be able to evaluate the integrals, the second is to recognize that the statement 'make I_{12} a constant' implies what in statistical mechanics is called a quenched system i.e. one must calculate the entropy for a given set of I_{ij} then average that entropy (the logarithm as in 2.2) over the initial and henceforward permanent values of the I_{ij}. This latter step is discussed in many papers and a particular method, the replica method, has been invented to handle it. It will take us too far away from topology to discuss this, so a simplified version will be offered later.

3. The evaluation of Gaussian invariants. Fick's equation (2.1) is like the Schrodinger equation and albeit "non relativistic" the evaluation of integrals of the Gauss type is like a field theory calculation. Such calculations started in the thirties but ran into divergence problems but in the forties it was realised that the divergence in some of the field theories was not an insuperable barrier to calculations for experiments measured the difference between the value of the field theoretic integrals when say

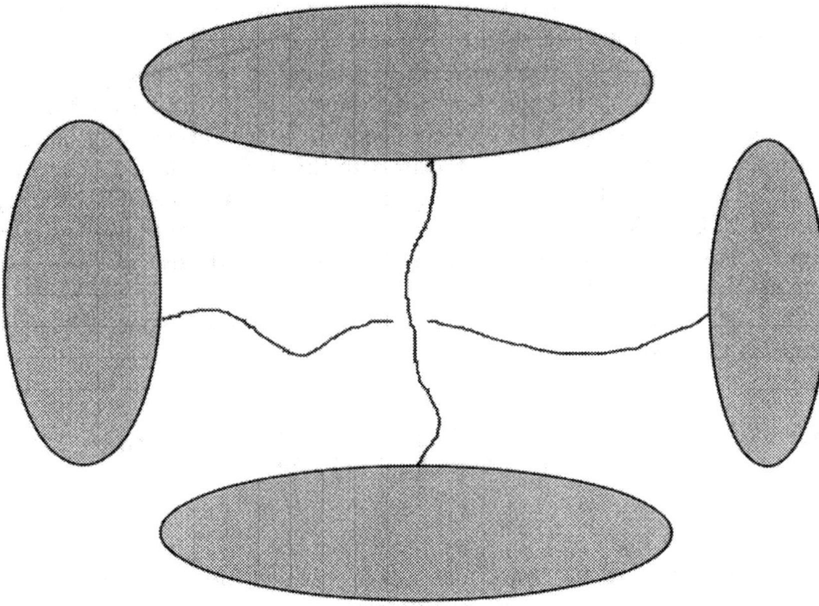

FIG. 1.

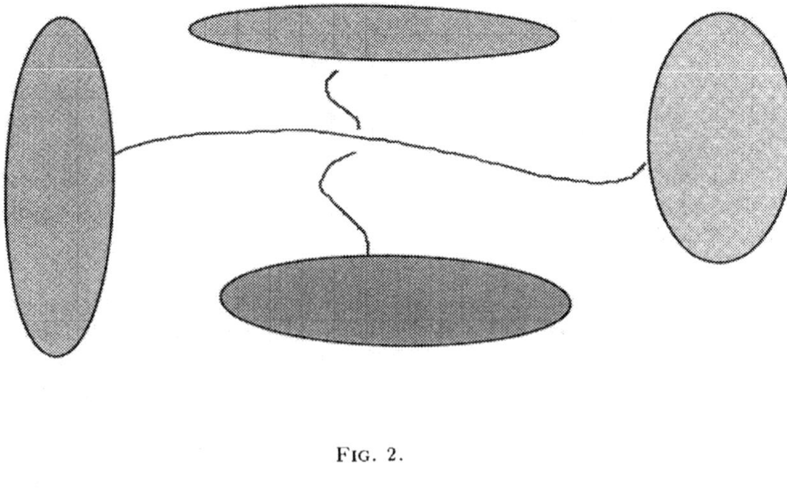

FIG. 2.

an external field is present. A typical integral might be

$$J(A) = \int_1^\infty \frac{dx}{x+A} \tag{3.1}$$

where experiment measures

$$J(A) - J(0) = \int_1^\infty \frac{dx}{x+A} - \int_1^\infty \frac{dx}{x} \tag{3.2}$$

$$= -\log(1+A). \tag{3.3}$$

This process, called renormalisation, is now fifty years old and so hallowed by time. Nevertheless, at a fundamental level one is faced by the fact that as A becomes large the integral becomes large, and if our present theories are inadequate at short distances and times, this inadequacy will show up at large A, i.e. at large external fields.

In polymers everything must be finite; the question we are addressing is whether one can discard finiteness guarantees and use Wiener type integrals to simplify calculations.

The first point is that the Gaussian invariant is the manifestation of a divergent but renormalisable field theory. The present author was able to prove this long ago[4] by showing that the Gaussian integral could be parametrised by a vector field $A_i(v)$ via a Lagrangian

$$\sum \int R^2 ds + \mu \int A \cdot R' ds + \lambda (A_k \times A_{-k}) \cdot k d^3 k \tag{3.4}$$

(Similar things have been done in high energy field theories recently by Chern and by Simons.) Since this point will appear soon further detail will not be given here; the argument below is self contained.

As a caution it should be remarked that it is possible to get wrong answers by renormalisation, for example if one calculates the osmotic pressure of a polymer solution the answer is

$$wc^2 + \int \log\left(1 + \frac{wc}{k^2}\right) d^3 k \tag{3.5}$$

where w is the interaction and c the concentration. This is renormised by removing a self interaction term $wc \int \frac{d^3k}{k^2}$ to leave a finite term[1]

$$wc^2 - b(wc)^{\frac{3}{2}} \tag{3.6}$$

If c becomes small, the fluctuation $c^{3/2}$ exceeds the mean c^2 and a negative osmotic pressure results which is physically impossible. If all the correct physical limitations to the theory are added, the theory becomes sensible. Crudely speaking, one just cuts off k at l^{-1} above and $(Nl)^{-1}$ from below. (In quantum field theories, relativistic invariance prohibits such cut offs.)

4. Calculation of the tube[6,7,8]**.** Firstly consider a crude argument. Suppose our polymers are swollen to a cylindrical shape of cross section πa^2 and then completely fill space[5]

$$\text{Then} \qquad \pi a^2 L = V$$

$$a = (c\pi)^{-\frac{1}{2}}. \tag{4.1}$$

But suppose that we argue (1) that calculations using Wiener measure can only give answers depending on $(L1)^{\frac{1}{2}}$. Then

$$\begin{aligned} a &= c^{\alpha} 1^{2\alpha+1} \\ &= V^{-\alpha} L^{\alpha} 1^{2\alpha+1} \end{aligned} \tag{4.2}$$

hence

$$2\alpha + 1 = \alpha$$

$$a = c^{-1} 1^{-1} \tag{4.3}$$

(Experimentally (9)a is available and lies between 1 and 1/2). These two results can be derived from the Gaussian integral where (4.1) results from differentiable chains and (4.3) from strictly Wiener chains i.e. continuous but not differentiable.

I will not go into the detail of the calculation here, but just outline it. One assumes there is an 'a' and evaluates it by considering the model of section (1) above i.e. one chain has fixed ends and moves amongst all the others which are frozen. Then consider our special chain at two remote times $R_1 R_2$ and crudely models $\prod_\alpha \delta(I_{1\alpha} - I_{2\alpha})$ where $I_{1\alpha}$ is the Gauss integral between 1 and another, α, by

$$\exp(-\int (R_1 - R_2)^2) a^{-4} l ds,$$

The entropy of the system is calculated, and in both cases of differentiable and non differentiable chains divergers, but is renormalisable.
Differentiable chains

$$S = La^{-2} + qcL \log a + S_0 \tag{4.4}$$

Wiener chains

$$S = La^{-2} + pca^{-1} + S_0 \tag{4.5}$$

where q, p are numerical factors and S_0 the rest of the entropy which is logarithmically divergent in (4.4) and a linear divergence in (4.5).

It is possible to extrapolate between these of course because the basic problem is always convergent.[10]

5. Conclusion. The difficulty with calculations like that of section 4 is that not only is the calculation tricky, but it is based on very crude topological arguments. However, it appears that the local nature of the chain is more important than the adequacy of the topological invariants.

Finally, it is perhaps worth remarking that dynamics offers a complete evasion of the topological problem but at the high price of needing solutions in time.[12] The equation of polymers moving in the presence of a fluid is

$$\left(\frac{\partial}{\partial t} - \sum_{\alpha,\beta} \int\int \frac{\partial}{\partial R(s_\alpha)} D(R(s_\alpha) - R(s_\beta)) \left(\frac{\partial}{\partial R(s_\beta)} + \frac{3\partial^2 R_\beta}{l\partial s_\beta^2}\right)\right) P = 0$$

(5.1)

where D is the Oseen tensor(roughly r^{-1}).

It is easy to show such chains cannot cross and so *all* topological invariants are the invariants of this equation. But I cant solve it.

REFERENCES

[1] Doi, M. & Edwards, S.F. 1986 *The theory of polymer dynamics*. Oxford University Press.
[2] de Gennes, P.G. 1979 Scaling concepts in polymer physics. Ithaca, New York: Cornell University Press.
[3] Flory, P.J. 1969 *Statistical mechanics of chain molecules*. New York. Interscience.
[4] Edwards, S.F. 1967 *Proc. Phys. Soc.* 92, 9.
[5] Edwards, S.F. 1968 *J. Phys.* A 1, 15.
[6] Edwards, S.F. & Evans, K. 1981 *J. Chem. Soc. Faraday Trans.* II. 77, 1913.
[7] Helfand, E. & Pearson, D.S. 1983 *J. Chem. Phys.* 79, 2054.
[8] Needs, R.J. & Edwards, S.F. 1983 *Macromolecules* 16, 1492.
[9] Isono, Y. et al. 1972 *Macromolecules* 5, 786.
[10] Edwards, S.F. 1988 *Proc. Roy. Soc. Lond.* A419, 221-234.
[11] Edwards, S F. 1982 *Proc. Roy Soc* A385, 267.

ENTROPIC EXPONENTS OF KNOTTED LATTICE POLYGONS

ENZO ORLANDINI*, E.J. JANSE VAN RENSBURG†, MARIA CARLA TESI‡, AND S.G. WHITTINGTON§

Abstract. Ring polymers in three dimensions can be knotted and the dependence of their critical behaviour on knot type is an open question. We study this problem for self-avoiding polygons on the simple cubic lattice using a novel grand-canonical Monte Carlo algorithm which is a combination of the BFACF algorithm and MMC sampling. We present numerical evidence that the entropic exponent depends on the knot type of the polygon. We conjecture that the exponent increases by unity for each additional factor in the knot factorization of the polygon.

1. Introduction. Polymer molecules in solution are typically very flexible objects which can be highly self-entangled, as well as being entangled with other molecules. It is important to understand and to characterize the extent of this entanglement complexity since it influences crystallization behaviour [1] as well as rheological properties [2]. If a ring closure reaction occurs the entanglement can be trapped as a knot in the resulting ring polymer, and some information about the entanglement complexity can be obtained from a study of the distribution of knots in the ring polymer. In addition, the presence of knots in closed circular DNA can give information about the mechanism of action of enzymes acting on the DNA molecule [3,4,5].

In the last few years some progress has been made towards answering some basic questions about knots, such as discovering the probability that a sufficiently long polymer is knotted [6,7] and the distribution of random knots [8]. However, the statistics of polymer rings with fixed knot type are less well understood. Quantities such as the average size of knotted polymer rings or their configurational entropy (i.e. the number of ways in which such rings can be embedded in space) could in principle depend on the knot type and such dependence can influence the mobility of long polymer chains such as DNA [4]. The question of how the size of a ring polymer depends on its knot type was first addressed by Janse van Rensburg and Whittington [9] by means of Monte Carlo methods. They showed that, although the dimensions of polygons are sensitive to the knot type, the critical exponent ν and the leading amplitude are independent of it. Subsequent studies confirmed the robustness of the critical exponent ν with

* CEA-Saclay, Service de Physique Théorique, F-91191 Gif-sur-Yvette Cedex, FRANCE.

† Department of Mathematics and Statistics, York University, Downsview, Ontario, Canada M3J 1P3.

‡ Université de Paris-Sud, Mathématiques Bat 425, 91405 Orsay Cedex, France.

§ Department of Chemistry, University of Toronto, Toronto, Ontario, Canada M5S 1A1.

respect to the change of the knot type while a debate is still open concerning the behaviour of the leading amplitude [10,11]. In this paper we will try instead to address the question of how the configurational properties of the ring polymer can be affected by the topological constraint of having a fixed knot type.

In order to investigate this problem we need a simple model which reflects the topological constraints (including the excluded volume effect) and the large scale properties of the polymer in solution. For a linear polymer in a good solvent it is well known that the self-avoiding walk model captures the essential physics of the excluded volume effect and faithfully reproduces the large scale metric properties, at least for large n. Since we are interested in the properties of ring polymers the corresponding model is a self-avoiding polygon i.e. an embedding of the circle graph in Z^3. In other words, an n-step self-avoiding polygon or *polygon* on the simple cubic lattice is an ordered sequence of points of Z^3 such that successive points are unit distant apart, all the points are distinct, and also the first and the last vertices are unit distant apart.

Let p_n be the number of distinct (up to translation) unrooted and undirected n-edge polygons, and let $p_n(k)$ be the corresponding number when the polygon is conditioned to have a particular knot type k. We shall write $k = \emptyset$ for the unknot and otherwise use the Alexander-Briggs notation so that a trefoil is 3_1, a figure eight is 4_1, etc. It is known [6,7] that

$$\lim_{n\to\infty} n^{-1}\log p_n(\emptyset) \equiv \kappa_o < \lim_{n\to\infty} n^{-1}\log p_n \equiv \kappa \tag{1.1}$$

so that unknots are exponentially rare in the set of all polygons. There is a similar result for any fixed knot type k [12]

$$\limsup_{n\to\infty} n^{-1}\log p_n(k) < \kappa \tag{1.2}$$

but there are important open questions about the relative numbers of polygons with different fixed knot types. It is easy to show that

$$\liminf_{n\to\infty} n^{-1}\log p_n(k) \geq \kappa_o \tag{1.3}$$

for any knot type k, and that

$$\limsup_{n\to\infty} n^{-1}\log p_n(l) \leq \liminf_{n\to\infty} n^{-1}\log p_n(k) \tag{1.4}$$

for any compound knot k which has l as a factor [13]. An important open question is whether the exponential growth rate depends on the knot type.

There are a few numerical results about the relative frequencies of lattice polygons with different knot types (see for instance [14]) and a detailed study of this question for Gaussian random polygons [15,16]. Using Monte

Carlo methods, Deguchi and Tsurusaki showed that the relative frequency of occurrence of each knot type (other than the unknot) increases as n increases, goes through a maximum, and then decreases. In fact (for lattice polygons) (1.2) implies that this decrease at large values of n must be exponential. The simpler knots have their maxima at smaller values of n and the location of the maximum generally moves to larger n as the knot increases in complexity [15,16].

For the set of all (unrooted) polygons, it is believed that the number of n-gons p_n has the following scaling form

$$p_n = An^{\alpha-3}\mu^n \left(1 + \frac{B}{n^\Delta} + \dots\right) \tag{1.5}$$

where $\mu = e^\kappa$ and α is the entropic exponent. The exponent Δ takes care of confluent corrections to scaling and, as α, is believed to be universal. The plausible extension of this to polygons with knot type k is

$$p_n(k) = A(k)n^{\alpha(k)-3}\mu(k)^n \left(1 + \frac{B(k)}{n^{\Delta(k)}} + \dots\right). \tag{1.6}$$

If this form is indeed correct then (1.2) implies that $\mu(k) < \mu$ for every k.

The primary aim of this work is to check the validity of eq. (1.6) and in particular to estimate the value of the entropic exponent (critical exponent), $\alpha(k)$, for some simple knots. The approach we follow is based on Markov chain Monte Carlo simulations. The algorithm we use is a novel algorithm which involves implementing the BFACF algorithm with MMC sampling. The BFACF algorithm consists of local moves which can change the length of the polygon but not the knot type. However, the algorithm suffers from very long autocorrelation times, which makes it unsuitable for a Monte Carlo study of very long polygons. To remedy this situation we implement the BFACF algorithm with a novel sampling method called Multiple Markov chains (MMC) [17,18].

Using this method we generate samples of polygons with some simple knot types and estimate their configurational entropy. From these results we estimate the corresponding critical exponents $\alpha(k)$. The plan of the paper is the following: In section 2 we introduce the BFACF algorithm and the MMC method. In section 3 we present the numerical results we obtain for different simple knot types and we give estimates of the entropic exponents. Section 4 is devoted to summary and discussion.

2. The BFACF algorithm and Multiple Markov chain sampling. The BFACF algorithm [19,20,21] involves sampling along a realization of a Markov chain defined on the set of polygons ω. It operates on ω by attempting one of the local deformations shown in figure 1 at a randomly selected location on the edges of ω.

In two dimensions the BFACF algorithm is known to be ergodic [22], whereas in three dimensions it is not ergodic, due to the possibility of

FIG. 1. *Elementary moves used in the BFACF algorithm*

knots [23]. In particular it has been proved that, for the simple cubic lattice the ergodicity classes of the algorithm are exactly the knot-types of the polygon [24]. This means that the set of polygons sampled in any realization is the set which has the same knot type as the initial polygon. This makes the BFACF method a convenient tool to study the influence of knots on the critical behaviour of polygons.

Unfortunately, the BFACF algorithm has very long autocorrelation times which leads to large statistical errors. The exponential autocorrelation time of this algorithm (which controls the relaxation of the Markov chain from an initial configuration to equilibrium) is infinite [25] and the integrated autocorrelation time, which controls the statistical error in measured quantities, is also very long.

To remedy this situation, algorithms that are combinations of BFACF moves and non-local moves have been proposed [9,26]. These algorithms are hybrids in which the non-local moves hopefully assure the rapid equilibration within subspaces of fixed N, while the local (BFACF) moves assure equilibration between different N.

However, if one wants to study critical behaviour of polygons *at fixed knot type* the approach described above presents a shortcoming: *The non-local moves do not preserve the topology of the polygon.* One way to deal with this problem is to check potential changes in knot type by calculating a topological invariant such as the Alexander polynomial at every successful non-local move [9]. This is computationally costly and has also the disadvantage that the Alexander polynomial is not a perfect invariant since distinct knots can have the same Alexander polynomial.

In this work we consider a different approach to speed up the BFACF algorithm in $d = 3$: instead of using additional (non-local) moves we try to increase the mobility of the BFACF algorithm by sampling along a set of Markov chains run in parallel.

This method has been introduced by Geyer [17] and then made suitable for simulating interacting self-avoiding walks with fixed length by Tesi et. al [18] (see also [27] for a more detailed analysis of the method).

We describe here a variant of this approach that we found to be suitable for simulations in the grand canonical (N varying) ensemble. The starting point is the single BFACF Markov chain whose limiting distribution $\pi(\omega)$ depends on the step fugacity K. For values of K close to the critical value K_c the Markov chain samples configurations with high values of N and the convergence of the BFACF algorithm is very slow. On the contrary for $K = K' << K_c$ the sampled configurations have small N and the convergence of the BFACF is fast. The idea is to select a set of values $K' = K_1 < K_2 < K_3 < \ldots < K_m = K$ to interpolate between K' and K, so that there is considerable overlap in the distributions at K_l and K_{l+1}.

The m Markov chains at $K_1, K_2, \ldots, K_m$ are evolved in parallel, and we allow the chains to interact (and possibly exchange conformations) as follows:

Select two neighboring chains at (say) fugacities K_j and K_{j+1} with uniform probability. A trial move is an attempt to swap the two current conformations of these chains. If $\pi_k(K)$ is the probability of the state k in the chain at fugacity K, and S_j and S_{j+1} are the current states in the j-th and $(j+1)$-th chain, then we accept the trial move (and swap S_j and S_{j+1}) with probability

$$r(S_j, S_{j+1}) = \min\left(1, \frac{\pi_{S_{j+1}}(K_j)\pi_{S_j}(K_{j+1})}{\pi_{S_j}(K_j)\pi_{S_{j+1}}(K_{j+1})}\right). \tag{2.1}$$

The whole process is itself a Markov chain, which we call the *composite Markov chain*. Since the underlying Markov chains are ergodic (for a fixed knot class), so is the composite Markov chain. Moreover the composite chain is in detailed balance since the "swap"-move as well as the moves in the underlying chain are. Consequently, the invariant limit distribution is the product distribution of separate Markov chains at the temperatures $K_1 < K_2 < \ldots < K_m$.

One can understand the advantage of this method by focusing on a particular K_j value. Every successful swap involving this chain and the chain K_{j+1} corresponds to a large change in the configuration at that value of K_j, so the correlation time at each K_j is markedly reduced. Swaps move the system (at a particular K_j) to new regions of configuration space and therefore tend to mitigate quasi-ergodic problems. Of course, some K_j values should be chosen so that convergence at those values is rapid. If the chains K_j and K_{j+1} are far apart the changes in the configuration are likely to be large but eq.(2.1) suggests that the bigger is the distance between the two selected chains the less likely it is that the swap will be accepted.

There are some practical considerations which require attention. First one has to decide on the set of K values. Clearly they should be sufficiently close together (and the value of m should be sufficiently large) that swapping occurs rather frequently, and K_1 should be small enough that convergence at that K value is rapid. Hence the gain in convergence rate is obtained at only marginal cost in computer time.

In Tables 1 and 2 we compare estimates of the integrated autocorrelation times (given in units of sampling, i.e. per 10^4 attempted BFACF moves in our tests) of the average polygon lengths, as a function of K, for two different cases: the pure BFACF algorithm and the BFACF algorithm with MMC sampling. In both cases the integrated autocorrelation times

TABLE 1

Integrated autocorrelation times for the mean length $\langle n\rangle$ of the UNKNOT. For each group of two columns the first column refers to the pure BFACF algorithm and the second column refers to the Multiple Markov chain algorithm with swaps between nearest neighboring chains. Each run has total length of 10^8 BFACF moves and the unit of sampling is 10^4 BFACF moves. The data which are not reliable are omitted and replaced in the table by dashed segments.

K	pure BFACF		BFACF + MMC	
	$\langle n\rangle$	$\tau(n)$	$\langle n\rangle$	$\tau(n)$
0.2030	26 ± 0.4	2.0 ± 0.1	26 ± 0.2	0.6 ± 0.03
0.2050	33 ± 0.8	4.8 ± 0.6	32 ± 0.3	0.6 ± 0.03
0.2070	42 ± 1.5	8.6 ± 1.4	42 ± 0.4	0.8 ± 0.05
0.2090	61 ± 3.4	20.3 ± 3.8	59 ± 0.8	1.2 ± 0.1
0.2100	71 ± 5.4	43.3 ± 15.7	76 ± 1.5	2.6 ± 0.5
0.2105	93 ± 8.5	52.9 ± 18.2	89 ± 1.7	2.3 ± 0.3
0.2110	106 ± 12.9	86.2 ± 27.2	105 ± 2.9	5.0 ± 1.2
0.2115	------	------	132 ± 3.8	5.5 ± 1.2
0.2120	------	------	167 ± 6.3	9.8 ± 2.8
0.2124	------	------	211 ± 9.3	14.4 ± 5.1
0.2128	------	------	291 ± 16.0	22.7 ± 10.5
0.2130	------	------	352 ± 24.0	33.3 ± 17.3

for the polygon length increases as K approaches the critical value K_c, as would be expected. However, it is clear that swapping leads to a significant decrease in the integrated autocorrelation times as K approaches the critical value.

3. Scaling arguments and entropic exponents. The BFACF algorithm realizes a Markov chain with limit distribution $\pi_k(\omega)$ given by

$$\pi_k(\omega) = \frac{1}{G_q}|\omega|^q K^{|\omega|}\chi(k(\omega), k) \tag{3.1}$$

where q and K are parameters which can be chosen to optimize the sampling, $|\omega|$ is the number of edges in the polygon ω, $k(\omega)$ is the knot type of ω, χ is an indicator function which is 1 if ω has the same knot type (k) as the first polygon in the realization of the Markov chain and zero otherwise. The normalization factor G_q is the grand partition function (generating function) of the problem. It depends on the fugacity K and on the knot type k and can be written as:

$$G_q(K, k) = \sum_n p_n(k) K^n n^q. \tag{3.2}$$

By inserting in eq. (3.2) the scaling behavior (1.6) for p_n we obtain for $G_q(K)$ the following asymptotic behaviour:

TABLE 2
As in table 1 but for the trefoil (knot 3_1).

K	pure BFACF		BFACF + MMC	
	$\langle n\rangle$	$\tau(n)$	$\langle n\rangle$	$\tau(n)$
0.2030	91 ± 1.2	5.8 ± 0.6	89 ± 0.5	1.1 ± 0.1
0.2050	103 ± 1.9	10.5 ± 1.9	104 ± 0.7	1.3 ± 0.1
0.2070	124 ± 3.0	17.3 ± 3.3	125 ± 0.9	1.5 ± 0.2
0.2090	161 ± 7.5	49.4 ± 13.9	161 ± 2.2	4.6 ± 0.8
0.2100	------	------	196 ± 4.0	9.6 ± 2.9
0.2105	------	------	216 ± 6.1	17.1 ± 6.0
0.2110	------	------	244 ± 6.9	16.4 ± 5.3
0.2115	------	------	284 ± 8.6	17.8 ± 6.1
0.2120	------	------	345 ± 11.2	17.6 ± 5.9
0.2124	------	------	434 ± 16.0	21.0 ± 8.7

$$G_q(K,k) \approx$$

$$A'(k)\,(1-\mu(k)K)^{2-\alpha(k)-q}\left[1+B'(k)\,(1-\mu(k)K)^{\Delta(k)}\cdots\right] \tag{3.3}$$

Hence, the mean number of edges in polygons of knot type k sampled at q and K is given by

$$\langle n(k)\rangle = K^{-1}\frac{\partial G_q(K)}{\partial K}$$

$$\approx \frac{[\alpha(k)+q-2]\mu(k)K}{1-K\mu(k)}\left(1-\frac{B(k)\Delta(k)[1-K\mu(k)]^{\Delta(k)}}{\alpha(k)+q-2}\right), \tag{3.4}$$

where we have made use of (3.3). In Table 3 we give estimates of $\langle n(k)\rangle$ as a function of K for runs carried out with $q=3$. In each case we used 16 parallel Markov chains and sampled every 10^5 attempted BFACF moves. The results are based on 80,000 sample points (at each value of K) for the unknot, 85,000 sample points for the trefoil, 95,000 sample points for the figure eight knot, 99500 sample points for 6_2 and 110,000 sample points for the square knot.

In our analysis of the data we assume that, given two knot types k_1 and k_2, $\mu(k_1)=\mu(k_2)=\mu(\emptyset)$. We know from (1.2) and (1.3) that $\mu(\emptyset)\le \mu(k)<\mu=e^{\kappa}$ for any knot k, and numerical evidence suggests that $\mu(\emptyset)$ is numerically very close to μ [14]. Hence any error introduced by this assumption will be numerically very small. Taking the ratio of the averages in (3.4) for two different knot types k_1 and k_2, and assuming also that Δ is independent of knot type, gives

$$\frac{\langle n(k_1)\rangle}{\langle n(k_2)\rangle} \approx \frac{\alpha(k_1)+q-2}{\alpha(k_2)+q-2}\left[1+c(1-K\mu(\emptyset))^{\Delta}\right]. \tag{3.5}$$

TABLE 3

Estimates of the mean number of edges as a function of knot type for various values of K when $q = 3$.

K	$\langle n(\emptyset)\rangle$	$\langle n(3_1)\rangle$	$\langle n(4_1)\rangle$	$\langle n(6_2)\rangle$	$\langle n(3_1\#3_1)\rangle$
0.2030	26.1 ± 0.1	89.8 ± 0.1	111.4 ± 0.1	148.7 ± 0.2	222.5 ± 0.3
0.2050	32.0 ± 0.1	104.1 ± 0.2	127.7 ± 0.2	170.2 ± 0.2	253.8 ± 0.4
0.2070	41.5 ± 0.2	126.1 ± 0.2	153.5 ± 0.2	203.1 ± 0.3	300.4 ± 0.5
0.2090	59.6 ± 0.2	165.9 ± 0.4	199.5 ± 0.4	260.6 ± 0.5	380.9 ± 1.0
0.2100	76.2 ± 0.3	200.8 ± 0.6	240.5 ± 0.6	311.0 ± 0.7	451.7 ± 1.5
0.2105	88.9 ± 0.4	227.9 ± 0.8	270.2 ± 0.8	348.5 ± 0.9	499.2 ± 1.9
0.2110	107.2 ± 0.5	264.0 ± 1.2	310.7 ± 1.1	397.6 ± 1.4	568.9 ± 2.9
0.2115	132.7 ± 0.9	318.1 ± 1.9	370.1 ± 2	470.7 ± 2.2	661.7 ± 5.0
0.2120	179.3 ± 1.5	408.3 ± 4.1	466.9 ± 4.5	586.9 ± 4.4	812.9 ± 7.6
0.2124	243.6 ± 2.7	533.9 ± 7.3	603.2 ± 9.5	741.3 ± 8.7	1019 ± 12
0.2128	379 ± 8	790 ± 17	883 ± 21	1033 ± 27	1394 ± 35
0.2130	522 ± 15	1067 ± 37	1170 ± 39	1325 ± 59	1857 ± 140
0.2131	636 ± 24	1293 ± 60	1404 ± 54	1541 ± 82	2286 ± 285
0.2132	830 ± 35	1658 ± 115	1795 ± 83	1884 ± 123	2853 ± 483

Since the exponent Δ is believed to have a value close to 0.5 [28,29], plotting $\langle n(k_1)\rangle/\langle n(k_2)\rangle$ against $(1 - K\mu(\emptyset))^{1/2}$ should give a curve which will become linear as K approaches $1/\mu(\emptyset)$ from below and have an intercept of $[\alpha(k_1) + q - 2]/[\alpha(k_2) + q - 2]$. In Figure 2 we show results, with $q = 3$, for $k_1 = 4_1$ and $k_2 = 3_1$, $k_1 = 3_1$ and $k_2 = \emptyset$, $k_1 = 3_1\#3_1$ and $k_2 = \emptyset$.

The fact that the curve for 4_1 and 3_1 is clearly approaching a value very close to unity strongly suggests that $\alpha(4_1) = \alpha(3_1)$. For the other two cases the intercept is difficult to determine precisely but we estimate that

$$\frac{\alpha(3_1) + 1}{\alpha(\emptyset) + 1} = 1.8 \pm 0.2 \tag{3.6}$$

and that

$$\frac{\alpha(3_1\#3_1) + 1}{\alpha(\emptyset) + 1} = 2.6 \pm 0.2 \tag{3.7}$$

In forming these estimates we have given considerable weight to the estimates at large values of K, corresponding to large values of $\langle n\rangle$.

In order to estimate the value of $\alpha(3_1)$ from this intercept we need a value for $\alpha(\emptyset)$. We have used our data to make a direct estimate of this quantity with the result $\alpha(\emptyset) = 0.27 \pm 0.03$ (three standard deviations) . The value of the intercept given in (3.6), together with our direct estimate (0.27), is consistent with $\alpha(3_1) = \alpha(\emptyset) + 1$. Similarly, the intercept in (3.7) is consistent with $\alpha(3_1\#3_1) = \alpha(\emptyset) + 2$.

Our results show that the exponent α clearly depends on the knot type of the polygon. On the other hands there is strong evidence that the

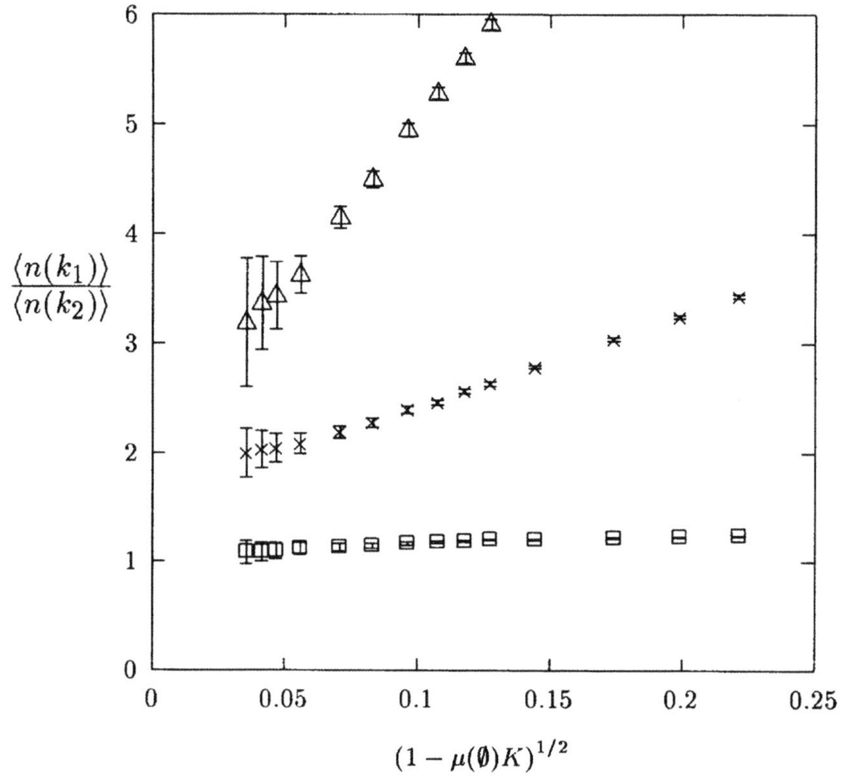

FIG. 2. *Plot of* $\langle n(k_1)\rangle/\langle n(k_2)\rangle$ *against* $(1-K\mu(\emptyset))^{1/2}$ *for* $k_1 = 4_1$ *and* $k_2 = 3_1$ *(□),* $k_1 = 3_1$ *and* $k_2 = \emptyset$ *(×),* $k_1 = 3_1\#3_1$ *and* $k_2 = \emptyset$ *(△).*

exponent ν, characterizing the radius of gyration, is independent of knot type [9]. This implies that the hyperscaling relation $d\nu = 2 - \alpha$ does not apply in three dimensions when the polygons are conditioned to be a particular knot type, and suggests that there is no analogue of the connection between polygons and the $O(n)$ model when the polygons have fixed knot type (see [30] for an introduction to hyperscaling in polymer rings). There are other examples known (e.g. c-animals) where the exponent controlling the growth of the number of objects depends on a topological restriction [31,32]. By analogy with the c-animals case, for which it is believed [33] that the entropic exponents for trees and animals are the same, one might expect that $\alpha(\emptyset)$ should be equal to α. The best estimate of α comes from a Monte Carlo estimate of ν together with hyperscaling [34], giving $\alpha = 0.237 \pm 0.005$ (three standard deviations). On the basis of our estimate for $\alpha(\emptyset)$ we cannot say anything conclusive about this point.

Our results for the unknot, trefoil, figure eight and square knots suggest that, for any knot k, the exponent $\alpha(k)$ might be given by

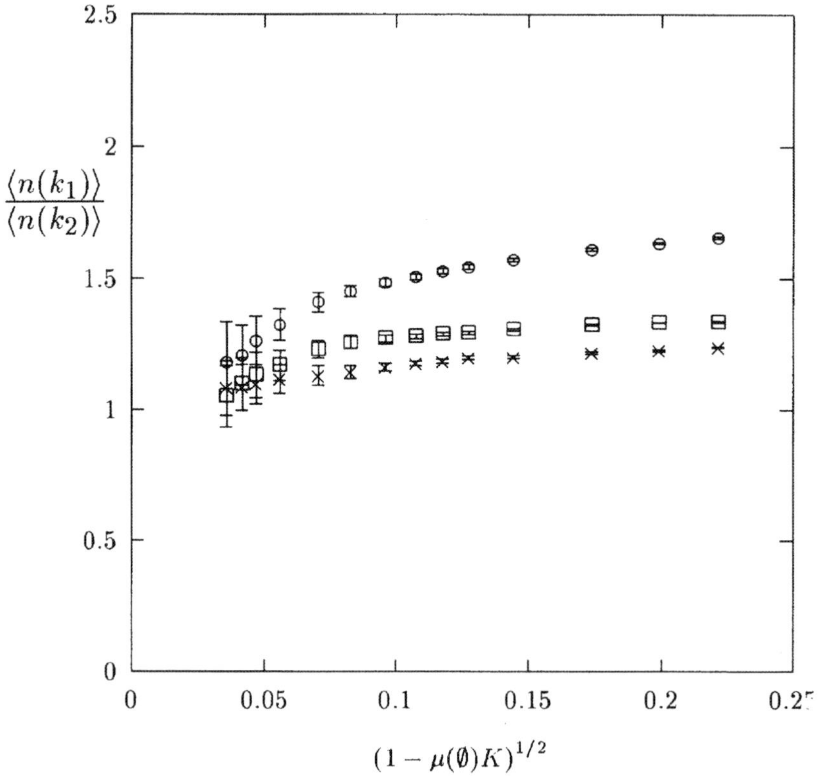

FIG. 3. *Plot of* $\langle n(k_1)\rangle/\langle n(k_2)\rangle$ *against* $(1-K\mu(\emptyset))^{1/2}$ *for* $k_1 = 4_1$ *and* $k_2 = 3_1$ *(×),* $k_1 = 6_2$ *and* $k_2 = 3_1$ *(□),* $k_1 = 6_2$ *and* $k_2 = 4_1$ *(O)*

$$\alpha(k) = \alpha(\emptyset) + N_f(k) \tag{3.8}$$

where $N_f(k)$ is the number of prime factors in the knot factorization of k. A particular consequence of eq. (3.8) is that all the prime knots should display the same critical behavior. In Figure 3 we plot the ratio $\langle k_1\rangle/\langle k_2\rangle$ for 3 different prime knots (3_1, 4_1 and 6_2) as a function of $(1-\mu(\emptyset)K)^{1/2}$. All the curves seem to approach unity supporting the above suggestion that the entropic exponent $\alpha(k)$ does not vary within the family of prime knots.

4. Discussion. Our main result is the suggestion that the entropic exponent α is independent of knot type for any non-trivial prime knot, and increases by unity for each factor in the knot factorization. How can we understand this heuristically, and what would be the consequences of such a result? One way to rationalize this result is to assume that most prime knots are close to being tight, in that they can be regarded as the concatenation of a small polygon of that knot type and a large unknotted polygon. If this is the case then one would expect that the concatenation could be carried out at a non-zero fraction of the edges of the unknotted polygon, so that for a prime knot τ

$$p_{n+m}(\tau) \approx \epsilon n p_n(\emptyset) \tag{4.1}$$

for some fixed m. If, in addition, the connective constant is independent of knot type, then this would imply that $\alpha(\tau) = \alpha(\emptyset) + 1$ for every prime knot τ. In fact a pattern theorem for unknotted polygons would immediately give (4.1) as an inequality.

How does this extend to composite knots? One might expect that each prime component behaves as a separate prime knot, so that the knotted polygons can be obtained by concatenating a large unknotted polygon with small polygons of each fixed prime knot type. If the number of components in the knot factorization is N_f, the analogue of (4.1) would be

$$p_{n+m}(\tau) \approx \binom{\epsilon n}{N_f} p_n(\emptyset) \tag{4.2}$$

Of course one would not expect the components to appear essentially separately for small values of n and there is some evidence for this in Figure 2. The curvature for the knot $3_1\#3_1$ in that figure is consistent with the two components not being separated for small n but separating as n increases.

If this scenario is correct one would expect that the amplitude of the radius of gyration might be independent of knot type, and that the differences might show themselves in correction to scaling terms. This is an interesting possibility which deserves further examination.

Acknowledgements. The authors wish to thank the Institute for Mathematics and its Applications (IMA) at University of Minnesota for its kind hospitality and for financial support. We are also pleased to acknowledge financial support from NSERC of Canada.

REFERENCES

[1] P.G. DE GENNES, *Tight knots*, Macromolecules **17** (1984), 703–704.

[2] S.F. EDWARDS, *Statistical mechanics with topological constraints: I*, Proc. Phys. Soc. **91** (1967), 513–519.

[3] S.A. WASSERMAN, J.M. DUNGAN AND N.R. COZZARELLI, *Discovery of a predicted DNA knot substantiates a model for site-specific recombination*, Science **229** (1985), 171–174.

[4] S.A. WASSERMAN AND N.R. COZZARELLI, *Biochemical topology: applications to DNA recombination and replication*, Science **232** (1986), 951.

[5] S.A. WASSERMAN AND N.R. COZZARELLI, *Supercoiled DNA-directed knotting by T_4 topoisomerase*, J. Biol. Chem. **266** (1991), 20567–20573.

[6] D.W. SUMNERS AND S.G. WHITTINGTON, *Knots in self-avoiding walks*, J. Phys. A: Math. Gen. **21** (1988), 1689–1694.

[7] N. PIPPENGER, *Knots in random walks*, Discrete Applied Math. **25** (1989), 273–278.

[8] V.V. Ribenkov, N.R. Cozzarelli and A.V. Vologodskii, *Probability of DNA knotting and the effective diameter of the DNA double helix*, Proc. Natl. Acad. Sci. U.S.A. **90** (1993), 5307–5311.
[9] E.J. Janse van Rensburg and S.G. Whittington, *The dimensions of knotted polygons*, J. Phys. A: Math. Gen. **24** (1991), 3935–3948.
[10] S. Quake, *Topological effects of knots in polymers*, Phys. Rev. Lett. **73** (1994), 3317–3320.
[11] A.Y. Grosberg, A. Feigel and Y. Rabin, *Flory-type theory of a knotted ring polymer*, Phy. Rev. E **54** (1996), 6618–6622.
[12] C.E. Soteros, D.W. Sumners and S.G. Whittington, *Entanglement complexity of graphs in Z^3*, Math. Proc. Camb. Phil. Soc. **111** (1992), 75–91.
[13] S.G. Whittington, *Topology of Polymers*, AMS Proc. Symp. Applied Math. **45** (1992), 73–95.
[14] E.J. Janse van Rensburg and S.G. Whittington, *The knot probability in lattice polygons*, J. Phys. A: Math. Gen. **23** (1990), 3573–3590.
[15] T. Deguchi and K. Tsurusaki, *Topology of closed random polygons*, J. Phys. Soc. Japan **62** (1993), 1411–1414.
[16] T. Deguchi and K. Tsurusaki, *Statistical Study of Random Knotting using Vasiliev Invariants*, in Random Knotting and Linking ed. K.C. Millett and D.W. Sumners (Singapore: World Scientific) (1994), 89–121.
[17] C.J. Geyer, *Markov chain Monte Carlo Maximum Likelihood*, in Computing Science and Statistics: Proceedings of the 23rd Symposium on the Interface, ed E.M. Keramidas (Interface Foundation) (1991), 156–163.
[18] M.C. Tesi, E.J. Janse van Rensburg, E. Orlandini and S.G. Whittington, *Monte Carlo study of the interacting self-avoiding walk model*, J. Stat. Phys. **82** (1996), 155–181.
[19] B. Berg and D. Foester, *Random paths and random surfaces on a digital computer*, Phys. Lett. **106B** (1981), 323–326.
[20] C. Aragao de Carvalho and S. Caracciolo, *A new Monte Carlo approach to the critical properties of self-avoiding random walks*, Journal de Physique **44** (1983), 323–331.
[21] C. Aragao de Carvalho, S. Caracciolo and J. Fröhlich, *Polymers and $g|\phi|^4$-theory in four dimensions*, Nucl. Phys. B [FS7] **215** (1983), 209–248.
[22] N. Madras (1986) (unpublished) and N. Madras and G. Slade, *The Self-avoiding Walk* (Boston: Birkhäuser) (1993) Chap. 9, 356–362.
[23] N. Madras and A.D. Sokal, *Nonergodicity of local length-conserving Monte Carlo algorithm for the self-avoiding walk*, J. Stat. Phys. **47** (1987), 573–595.
[24] E.J. Janse van Rensburg and S.G. Whittington, *The BFACF algorithm and knotted polygons*, J. Phys. A: Math. Gen. **24** (1991), 5553–5567.
[25] A.D. Sokal and L.E. Thomas, *Exponential convergence to equilibrium for a class of random walk models*, J. Stat. Phys. **54** (1989), 797–828.
[26] S. Caracciolo, A. Pellissetto and A.D. Sokal, *Nonlocal Monte Carlo algorithm for self-avoiding walks with fixed endpoints*, J. Stat. Phys (1990), 1–53.
[27] E. Orlandini, *Monte Carlo study of polymer systems by Multiple Markov Chain method*, to appear in "Numerical methods for Polymeric Systems", ed. S. Whittington, IMA Volumes in Mathematics and its Applications, (Springer Verlag 1998).
[28] J.C. Le Guillou and J. Zinn-Justin, *Critical exponents from field theory*, Phys. Rev. B **21** (1980), 3976–3998.
[29] J.C. Le Guillou and J. Zinn-Justin, *Accurate critical exponents from field theory*, Journal de Physique **50** (1989), 1365–1370.
[30] B. Duplantier, *Hyperscaling for polymer rings*, Nucl. Phys. B **430** (1994), 489–533.
[31] C.E. Soteros and S.G. Whittington, *Critical exponents for lattice animals with fixed cyclomatic index*, J. Phys. A: Math. Gen. **21** (1988), 2187–2193.

[32] D. ZHAO, Y. WU AND T. LOOKMAN, *On the critical exponent ν for c-animals*, J. Phys A: Math. Gen. **25** (1992), L1187–L1190.
[33] T.C. LUBENSKY AND J. ISAACSON, *Statistics of lattice animals and dilute branched polymers*, Phys. Rev. **A 20** (1979), 2130–2146.
[34] B. LI, N. MADRAS AND A.D. SOKAL, *Critical exponents, hyperscaling and universal amplitude ratios for two- and three-dimensional self-avoiding walks*, J. Stat. Phys. **80** (1995), 661–754.

THE TORSION OF THREE-DIMENSIONAL RANDOM WALK*

MIKELIS G. BICKIS†

Abstract. Measures of torsion for the sample paths of three-dimensional random walk are defined. Their asymptotic distributions are shown to be normal both in the case of unrestricted random walk, and in the case of non-reversal random walk.

MSC 1991. 60J15, 82D60, 82B41, 60F05.

Consider a simple random walk on the three-dimensional integer lattice — i.e., a stochastic process in discrete time having states in Z^3 and independent increments, the increments for one time step being uniformly distributed over the unit vectors in the six coordinate directions. A sample path of such a process defines a piecewise linear curve in space, and one can ask distributional questions about geometrical properties of such a curve. In this paper we will find the asymptotic distribution of the total torsion and total absolute torsion of such curves. These quantities measure the extent to which a random walk twists in space. Such twists are of interest in the statistical mechanics of polymers. A linear polymer can be modeled as a self-avoiding walk, and a ring polymer can be modeled by a self-avoiding polygon. The distributional properties of self-avoiding paths are too complicated to study analytically but can only be handled using Monte Carlo simulation (Tesi, 1996). The purpose of this note is to study the asymptotic behaviour for unconstrained random walks to see to what extent they resemble or differ from those satisfying self-avoidance properties.

1. Definition of torsion. The torsion of a smooth space curve K is typically defined as follows (see e.g. Millman & Parker, 1977). Let $\mathbf{y} : [0, L] \to R^3$ be a parametrization for the curve in terms of arc length (so that the distance along the curve from $\mathbf{y}(a)$ to $\mathbf{y}(b)$ is $b - a$. Then $\mathbf{T}(t) = \mathbf{y}'(t)$ is a unit vector in the tangent space to K at $\mathbf{y}(t)$. The normal to the curve at $\mathbf{y}(t)$ is defined as $\mathbf{N}(t) = \mathbf{T}'(t)/\kappa(t)$, where $\kappa(t) = ||\mathbf{T}'(t)||$ is the curvature. Wherever $\kappa \neq 0$, $\mathbf{N}$ will be a unit vector orthogonal to $\mathbf{T}$, and $\mathbf{T}$ and $\mathbf{N}$ will span the osculating plane to the curve. The normal to this plane is defined by the binormal $\mathbf{B} = \mathbf{T} \times \mathbf{N}$. The torsion at $\mathbf{y}(t)$ is then defined as $\tau(t) = -\mathbf{B}'(t) \cdot \mathbf{N}(t)$.

The torsion quantifies the extent to which the curve is moving away from its osculating plane: A plane curve has torsion 0. The total torsion is

---

* This research was done while the authour was in residence at the Institute of Mathematics and its Applications, University of Minnesota, Minneapolis MN, U.S.A..

† Department of Mathematics and Statistics, University of Saskatchewan, Saskatoon, SK S7N 5E6, Canada.

computed by integrating τ over the curve, giving a measure of how far it has twisted. A helix has constant torsion, which is positive if the helix is right-handed. Since τ is a signed quantity, the total torsion does not indicate how far the curve is from a plane curve, since, for example, a curve consisting of equal segments of a left-handed and a right-handed helix will have zero total torsion. Also, any closed curve inscribed on the surface of a sphere will have zero total torsion (Millman and Parker, 1977; p. 170). A measure of deviation from planarity can be obtained by integrating $|\tau|$ or τ^2.

The preceding discussion makes sense only for smooth space curves. The torsion cannot be defined this way unless $\mathbf{y}$ is at least three times differentiable, and even then will be undefined wherever the curvature is zero. Nevertheless, concepts of twisting and planarity make sense for space curves which may not be differentiable, or for piecewise linear curves which have zero curvature almost everywhere. We will therefore use a more general definition of torsion, following Alexander and Reshetnyak (1989).

It can be shown that the torsion as defined above is equal to the signed volume of the parallelepiped spanned by $\mathbf{T}$, $\mathbf{N}$, and $\mathbf{N}'$, considered to be positive where these vectors form a right-handed system. (If we imagine being seated on the curve facing the direction $\mathbf{T}$ with $\mathbf{N}$ pointing up, then positive torsion corresponds to $\mathbf{N}'$ pointing to the right.)

In this paper we will be considering random walk on the three-dimensional integer lattice. The position of the curve at the kth step is given by

$$\mathbf{S}_k = \sum_{i=1}^{k} \mathbf{X}_i, \tag{1.1}$$

where the $\mathbf{X}_i$ are i.i.d. random unit vectors. We will define τ_k, the torsion at the kth step, as the signed volume of the cube defined by $\mathbf{X}_{k^*}, \mathbf{X}_k, \mathbf{X}_{k+1}$, where $k^* = \max\{m < k | \mathbf{X}_m \perp \mathbf{X}_k\}$. The torsion will be ± 1 according as these vectors form a right-handed or left-handed system, and will be zero if they are linearly dependent. This definition is the discrete analogue for cubic lattice paths of the standard definition for smooth space curves.

2. Torsion for random walks. Our definition of torsion is not a local property, but depends on the recent history of the path. The "torsion process" can be made Markovian, however, by reparametrization. Let

$$(D_i, U_i) \quad i = 1, 2, \dots \tag{2.1}$$

be a sequence of i.i.d. random pairs, where D_i is uniformly distributed over the set $\{1, 2, 3\}$, U_i is uniformly distributed over $\{-1, 1\}$, and D_i and U_i are independent. Then the random walk (1.1) can be defined in terms of the sequence (2.1) by

$$\mathbf{X}_i = U_i \mathbf{e}_{D_i},$$

where $\mathbf{e}_1$, $\mathbf{e}_2$, $\mathbf{e}_3$ are the natural basis vectors in R^3. D_i indicates the dimension in which the walk proceeds, and U_i determines whether the step will be in the negative or positive direction.

Let $\alpha = \min\{k : D_k \neq D_1\}$. Define a process $\mathbf{R}_k = (R_k^1, R_k^2, R_k^3, R_k^4)$ as follows

$$
\begin{aligned}
\mathbf{R}_0 &= (D_1, D_\alpha, U_{\alpha-1}, U_\alpha) \\
\mathbf{R}_k &= \begin{cases} (R_{k-1}^1, R_{k-1}^2, R_{k-1}^3, U_{\alpha+k}) & \text{if } D_{\alpha+k} = R_{k-1}^2 \\ (R_{k-1}^2, D_{\alpha+k}, R_{k-1}^4, U_{\alpha+k}) & \text{if } D_{\alpha+k} \neq R_{k-1}^2 \end{cases}
\end{aligned}
\tag{2.2}
$$

Conditioning on the value of α, the process is a Markov chain with 24 states defined by quadruples (a, b, s, t) such that $a, b \in \{1, 2, 3\}$, $a \neq b$, and $s, t \in \{-1, 1\}$. The first and third components of $\mathbf{R}_k$ "remember" what the preceding direction of motion was, and thus allow one to define the plane spanned by the last part of the lattice path. From a state (a, b, s, t) the process will move to one of the six states $(a, b, s, \pm 1)$, $(b, a, t, \pm 1)$, $(b, c, t, \pm 1)$ (where $c \neq a, b$) with equal probabilities, and cannot move in one step to any other state.

The torsion of the random walk can now be defined in terms of the process (2.2). The torsion will be non-zero at precisely those steps where $\mathbf{R}_k = (a, b, s, t)$ and $\mathbf{R}_{k+1} = (b, c, t, u)$ with $a \neq c$. For those steps the torsion will be $stu\,\pi(a, b, c)$ where $\pi(a, b, c)$ is the parity of the permutation (a, b, c) of $(1, 2, 3)$.

A renewal process can now be embedded in the Markov chain (2.2). Let V_1 be the first k such that R_k^1, R_k^2 and R_{k+1}^2 are all distinct. Then the torsion will be zero at all steps before V_1, and will be ± 1 at the V_1st step. V_1 represents the last step for which the sample path remains in a plane. The distribution of V_1 will be geometric with expected value 3, since at every step there is a probability of $1/3$ of leaving the plane. Once the path leaves the plane, a new plane is defined by the V_1st and $(V_1 + 1)$st steps, and we can define V_2 as the sojourn time of the path in this new plane. Similarly one can define V_m as the sojourn time in the mth plane. The total torsion of a path of length n is then

$$
T_n = \sum_{i:\alpha+V_i<n} R_{V_i}^3 R_{V_i}^4 R_{V_i+1}^4 \,\pi(R_{V_i}^1, R_{V_i}^2, R_{V_i+1}^2)
\tag{2.3}
$$

A sequence of Bernoulli trials can also be embedded in the Markov process (2.2). Define a "success" as a transition from (a, b, s, t) to $(b, c, t, \pm 1)$ where $a \neq c$—i.e., a step in which the current osculating plane is abandoned. The probability of such a transition is $1/3$, and does not depend on the current state of the chain. It is precisely at these steps that the torsion is different from zero. Moreover, the torsion will be ± 1 with equal probabilities and independent of any other aspects of the transition. Let M

be the number of successes in n steps. Then the total torsion (2.3) has the same distribution as

$$T_n^* = \sum_{j=1}^{M} Z_j \tag{2.4}$$

where the Z_j are i.i.d. uniformly distributed on $\{-1, 1\}$. The total absolute torsion, on the other hand, is just equal to M.

The number of successes M does not quite have a binomial distribution since there are end effects. The number of trials is not n but $n - \alpha$. It is tedious but elementary to derive that the expected number of trials for an n-step walk is $n - 5/2 + 1/(2 \times 3^{n-2})$. The variance for the number of trials can similarly be calculated to be $3/4 + O(n\,3^{-n})$.

It then follows that

$$\begin{aligned} E(M) &= \frac{n}{3} - \frac{5}{6} + O(3^{-n}) \\ V(M) &= E(V(M|\alpha)) + V(E(M|\alpha)) \\ &= \frac{2}{9} E(n - \alpha) + \frac{1}{9} V(n - \alpha) \\ &= \frac{2}{9} n - \frac{17}{36} + O(n\,3^{-n}). \end{aligned}$$

From these relations it follows that the total torsion has expected value zero and asymptotic variance $n/3$, whereas the total absolute torsion has asymptotic mean $n/3$ and asymptotic variance $2n/9$.

The asymptotic normality of these quantities can also be established. For this we require a central limit theorem for the sum of a random number of random variables. The following is taken from Feller (1971) p. 265.

THEOREM 2.1. *Let $X_1, X_2, \ldots$ be i.i.d. random variables with zero expectation and unit variance, and $N_1, N_2, \ldots$ be positive integer valued random variables, independent of the X_i's, such that $N_n/n \to 1$ in probability. Then*

$$n^{-\frac{1}{2}} \sum_{i=1}^{N_n} X_i$$

converges in distribution to the standard normal distribution.

Putting

$$X_i = \begin{cases} \sqrt{2} & \text{for a success, and} \\ -1/\sqrt{2} & \text{for a failure} \end{cases}$$

with $N_n = n - \alpha$ establishes the asymptotic normality of M and hence of the total absolute torsion. Putting $X_i = Z_i$ and $N_n = M_{3n}$ establishes the asymptotic normality of the total torsion.

We have thus proved the following theorems:

THEOREM 2.2. *The total torsion of a three dimensional random walk of n steps is asymptotically normal with mean zero and variance $n/3$.*

THEOREM 2.3. *The total absolute torsion of a three dimensional random walk of n steps is asymptotically normal with mean $n/3$ and variance $2n/9$.*

3. Restricted walks. The distribution of torsion of a random walk is of interest in studying certain properties of polymers (Tesi, 1996). A random walk, however, is not a good model for polymers for it can self-intersect. A *self-avoiding walk* (Madras and Slade, 1993) is a better model, but is intractable analytically because the self-avoidance property induces a complicated dependency structure among the steps. We can come somewhat closer to a self-avoiding walk by at least excluding reversals from the permissible steps. That is, at each step of the walk, only five of the six coordinate directions are acceptable. We forbid the walk from returning to its immediately preceding position. Such a restricted walk also avoids an unsatisfying aspect of the torsion as defined above. If reversals are allowed, then one can construct sample paths which look like right-handed helices except that a reversal (i.e., two steps in opposite directions that cancel each other) is attached at each turn. The way the torsion is defined, these reversals have the effect of making the torsion everywhere negative, even though the general appearance of the helix is right-handed.

The derivations in the preceding section are easily modified to deal with the case of excluded reversals. If all five coordinate directions are given equal probability, then the probability of a "success" increases from $1/3$ to $2/5$, and the expected number of trials becomes $n - 9/4 + 1/(4 \times 5^{n-2})$. These lead to the following theorems:

THEOREM 3.1. *The total torsion of a three dimensional random walk with forbidden reversals and n steps is asymptotically normal with mean zero and variance $2n/5$.*

THEOREM 3.2. *The total absolute torsion of a three dimensional random walk with forbidden reversals and n steps is asymptotically normal with mean $2n/5$ and variance $6n/25$.*

One can only speculate about how close these asymptotic formulas are to the case of true self-avoiding walk. It should be noted that even the partial self-avoidance property of forbidden reversals has the effect of increasing the torsion of sample paths. It is reasonable to imagine that true self-avoidance would further increase the tendency of sample paths to twist. Using Monte Carlo methods, Tesi (1996) observed approximate normality of torsion for self-avoiding polygons (self-avoiding walks that return to the origin). She found empirically that $V(T_n)/n \approx 0.45 \pm .01$. This value, somewhat higher than the 0.4 predicted for random walks without reversals, supports the conjecture that self-avoidance increases the tendency for torsion.

REFERENCES

[1] Alexandrov, A.D. and Reshetnyak, Yu.G., *General Theory of Irregular Curves*, Dordrecht: Kluwer (1989).

[2] Feller, W., *An Introduction to Probability Theory and Its Applications*, **II**, 2nd ed. New York: Wiley (1971).

[3] Madras, N. and Slade, G., *The Self-Avoiding Walk*, Boston: Birkhäuser (1993).

[4] Millman, R.S. and Parker, G.D., *Elements of Differential Geometry*, Englewood Cliffs: Prentice-Hall (1977).

[5] Tesi, M.C., *Geometrical Entanglement in Lattice Models of Linear Polymers: Torsion and Writhe*, IMA Volumes in Mathematics and its Applications, "Numerical Methods for Polymeric Systems", Editor: Stuart Whittington, Vol. 102, Springer-Verlag New York, Inc. (1997).

SELF-REPELLING KNOTS AND LOCAL ENERGY MINIMA

LOUIS H. KAUFFMAN*, MILANA HUANG†, AND ROBERT P. GRESZCZUK‡

Abstract. This paper discusses properties of simulations of electrically self-repelling knots. In particular, we give an example of two configurations of the Kinoshita-Terasaka knot (of Alexander polynomial one) that appear to be distinct local energy minima for both the electrical simulation and for the Simon energy.

1. Introduction. In this paper we will describe a program for exploring charged and self-repelling knots on the computer and some specific experiments that we have done with this program. We will spend a short time describing our previous paper [4] in which we performed unknotting experiments using simulated annealing. In the present paper we have restricted ourselves to a simulation of an electrically charged loop in space and have asked the program some specific questions about its behaviour on certain knots.

The paper is organized in three sections beyond the introduction. The second section describes our program and its basic organization. The third section describes our experiment with the Kinoshita-Terasaka knot. We give an example of two configurations of the Kinoshita-Terasaka knot that appear to be distinct local energy minima for both our electrical simulation and for the Simon energy. The last section describes conclusions and problems for further research.

Acknowledgement. It gives the first author pleasure to thank the National Science Foundation for support of this research under NSF Grant DMS-2528707.

2. Basics of self-repulsion. First we give a description of the computer program: This program was written by Milana Huang, with advice on knot theory from Lou Kauffman and advice on simulated annealing from Robert Greszczuk. The basis of the program is a simulation of (possibly) knotted loops in three dimensional space where the loop is composed of straight segments of varying length. Each segment is modelled with a spring constant, and its length varies according to the forces on the springs computed according to Hooke's law (i.e. the force is directly proportional to the length of the segment). The (simulated) electrical force between segments is inversely proportional to the squared distance between the seg-

* Department of Mathematics, Statistics and Computer Science, University of Illinois at Chicago, 851 South Morgan Street, Chicago, IL 60607-7045.

† Electronic Visualization Laboratory, Department of EECS, University of Illinois at Chicago, Chicago, IL 60607-7045.

‡ Department of Radiology, University of Chicago, Chicago, IL 60637.

ments with a fixed constant of proportionality for each pair of segments (corresponding to the products of the charges on the two segments- i.e. we have a constant amount of charge per segment). We take the distance between two segments to be the *minimal* pointwise distance between them. This force can be computed by analytic geometry. It has the advantage that the electrical force between segments that are very close rises sharply. This enables the program to keep the loop from passing through itself. The total force on a given segment of the loop is the sum of the forces between this segment and all the other segments in the loop. The forces are directed between the centers of the segments. (Note that we compute the minimal distance, but take the force between the centers. This is useful computationally, but is physically unreal unless the segments are small in length.) In this way, we obtain a simulation of a springy knotted charged loop, and can observe its behaviour over time.

A program, *Ming*, has been written by Ying-Quing Wu at the University of Iowa [7]. Ming works with the energy functional defined by John Simon [2] (See also [6]). Simon's energy is inversely proportional to the squared distance between midpoints of segments. Hence the Simon energy does not correspond to the classical physical electrical force. Nevertheless, there is qualitative correspondence between Ming (using Simon energy) and our program (using electrical repulsion), as we shall comment on below.

3. A knot with distinct local minima. In our previous paper [4] we studied the unknotting abilities of our program and its enhanced unknotting capabilities using simulated annealing. The program has some limitations in its ability to unknot loops that are topologically unknotted. These limitations are described in our paper [4]. However, when coupled with a simulated annealing process our program is quite successful in unknotting quite complicated configurations and there does not seem to be any limitation, in principle, to its unknotting capability.

In this paper, we present examples of knots that appear to have distinct local energy minima, in the sense of an empirically stable end state to the process initiated by the dynamics of the program. The example we have chosen to study is based on the knot *KT,* the Kinoshita-Terasaka Knot. This knot, shown in Figure 1, is famous as the one of the smallest knots that is undetectable by the classical Alexander polynomial [1], [5]. In Figure 1 we show a 12 crossing version of KT, labelled $KT12$ and we illustrate an intermediate version of KT obtained from $KT12$ by swinging an arc by ambient isotopy. It is easy to then swing an arc in the intermediate to obtain the 11 crossing version of KT shown in Figure 2, labelled there $KT11$. We refer to these two initial configurations for KT as $KT11$ and $KT12$. $KT11$ has an 11 crossing diagram as shown in Figure 2, while $KT12$ has a 12 crossing diagram, as shown in Figure 1. From the point of view

KT12

KT-INTERMEDIATE

FIG. 1.

FIG. 2.

of a knot theorist, the 12 crossing configuration is quite interesting, as it shows that the knot KT is a ***ribbon knot***, and consequently KT bounds a smooth disk in upper four space.

It is an interesting exercise to see how to deform $KT12$ to $KT11$. In Figures 1 and 2 we have given the reader sufficient information to puzzle this out, with the help of the diagram KT-Intermediate in Figure 1.

We were skeptical whether our program would be able to find a deformation from $KT11$ to $KT12$ and so we performed the experiment. The initial input to such an experiment is either a file of endpoints for the segments of the embedded loop, or a diagram of the knot drawn with mouse coupled with mouse clicks to choose under and over crossings. We chose the latter method to input $KT11$ and $KT12$. Running the program on these initial data led to experimentally stable configurations that are visually distinct (These configurations also have nearly the same Simon energy [2], as can be computed using our program or the program Ming [7]). Thus we arrive at the conjecture that $KT11$ and $KT12$ produce distinct energy minima for the simulated electrical repulsion. In Figures 3,4,5,6 we have illustrated the computer pictures of the initial and final states for $KT11$ and $KT12$ with $KT11$ treated in Figures 3 and 4, $KT12$ treated in Figures 5 and 6.

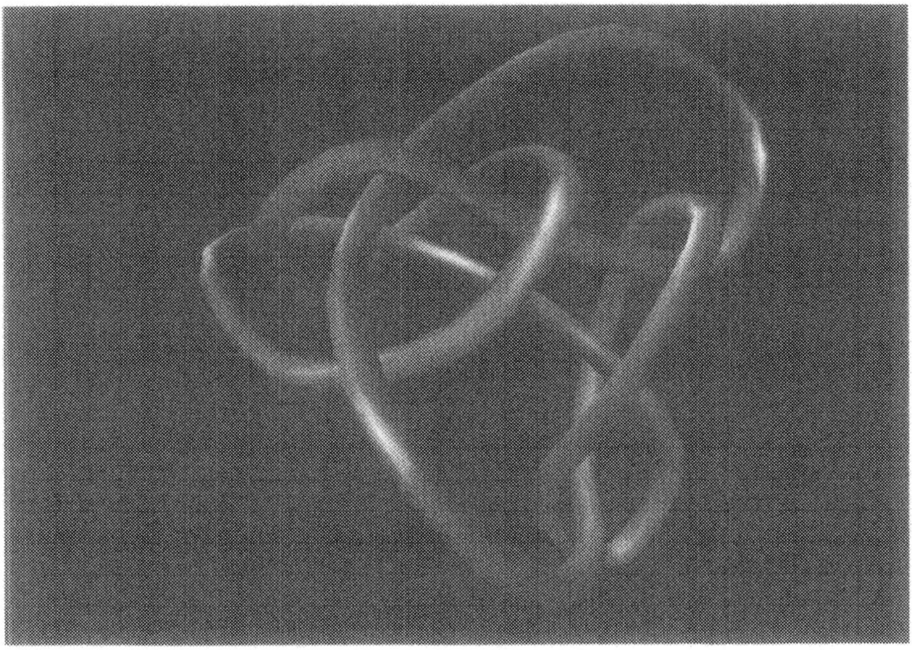

FIG. 3.

FIG. 4.

FIG. 5.

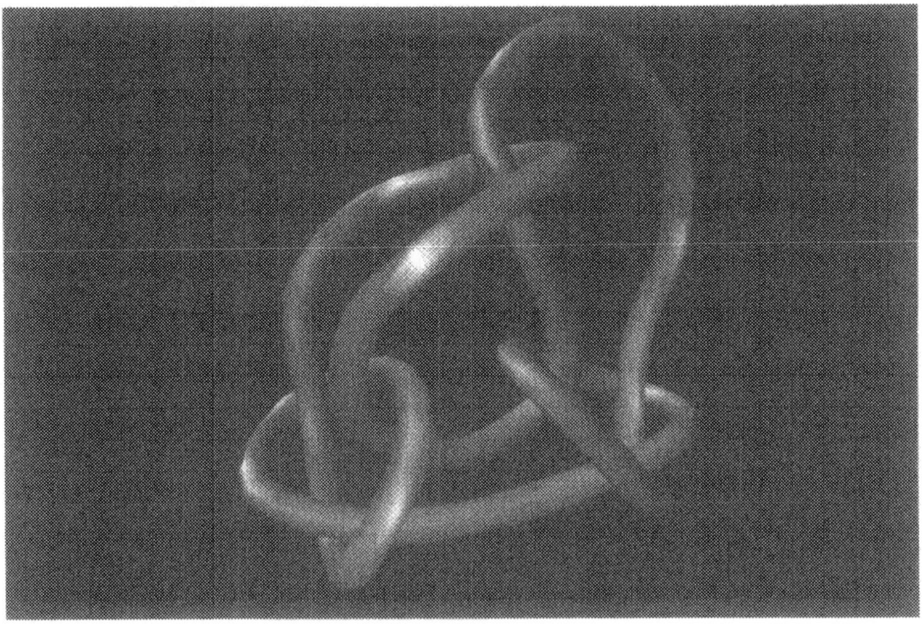

FIG. 6.

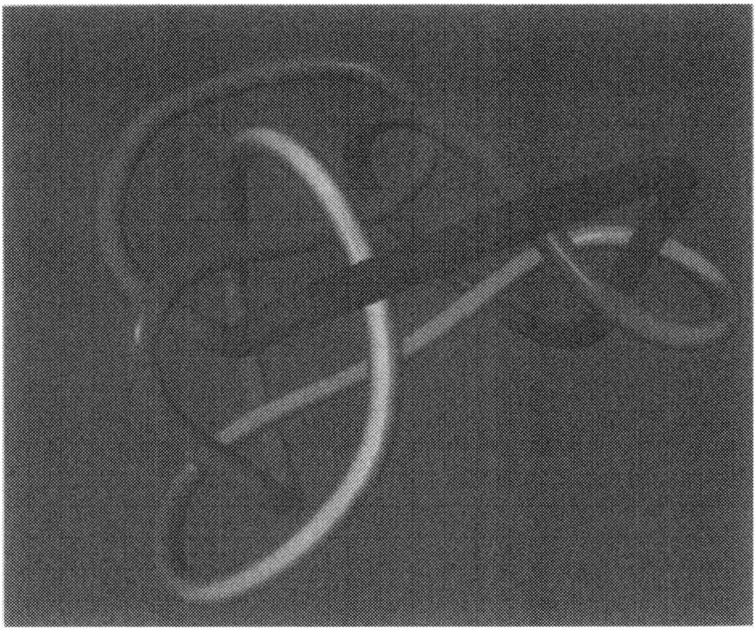

FIG. 7.

There are, of course, many experiments of this type that one can perform with our program. We were also interested to see if our results would be the same if we used the Simon energy minimization routine that is instantiated in the program Ming. Our files are exactly compatible with the the input for Ming, so it was possible to try the output for $KT11$ and $KT12$ on Ming. The results are substantially the same. $KT11$ and $KT12$ appear to produce distinct energy minima. In Figures 7 and 8 we show corresponding pictures of final states produced by Ming.

4. Conclusions. The conclusions of this paper are experimental, based on the actions of our program and of the program Ming. We would very much like to be able to prove rigorously that the Kinoshita-Terasaka knot does indeed have distinct local energy minima both for classical electrical repulsion and for the Simon energy. It is possible that such a proof can be constructed by a combination of analytic and computer work. Such a proof is one hope for the future.

In general it is quite stimulating to work with a program of this type. Flat data from knot diagrams becomes translated, via self-repulsion, into palpable three dimensional geometry related to the topology of the knot. It will require much more experimentation before the full richness of this relationship between geometry and topology is uncovered.

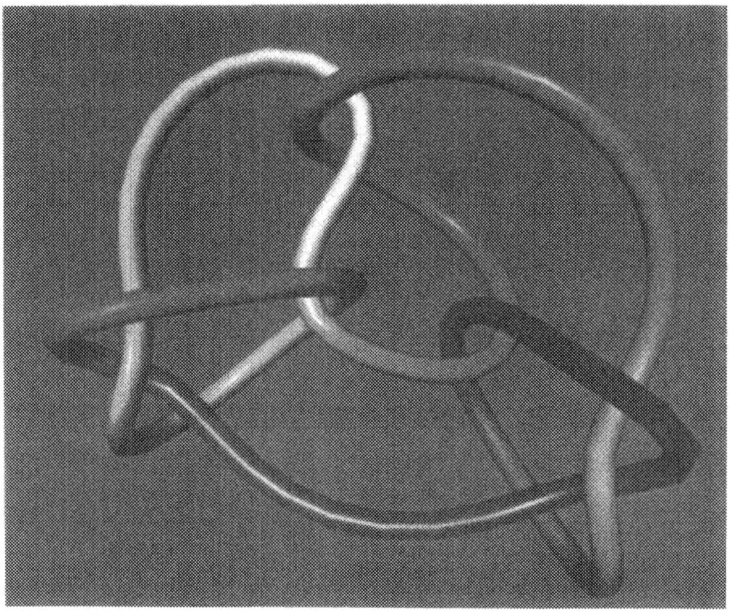

FIG. 8.

REFERENCES

[1] J.W. ALEXANDER, *Topological invariants of knots and links*, *Trans. Amer. Math. Soc.*, Vol. 20 (1923), 275–306.

[2] G. BUCK AND J. SIMON, *Knots as dynamical systems*, *Topology Appl.*, 51, 1993.

[3] MILANA HUANG, *Untangling Knots by Energy Minimization with Simulated Annealing*, Masters Thesis, Electrical Engineering and Computer Science, University of Illinois at Chicago, 1996.

[4] MILANA HUANG, ROBERT P. GRZESZCZUK AND LOUIS H. KAUFFMAN, *Untangling knots by stochastic energy optimization*, *to appear in the Proceedings of Visualization '96.*

[5] L.H. KAUFFMAN, *Knots and Physics*, World Scientific Publishing Co., 1991, 1994.

[6] R.B. KUSNER AND J.M. SULLIVAN, *Mobius energies for knots and links, surfaces and submanifolds*, Technical report, Geometry Center Research Report GCG64, University of Minnesota, 1994.

[7] YING-QUING WU, *Silicon Graphics Machine Program — Ming*, University of Iowa, 1996.

PROPERTIES OF KNOT ENERGIES

Y. DIAO*, C. ERNST†, AND E.J. JANSE VAN RENSBURG‡

Abstract. Energy functions on knots are continuous and scale-invariant functions defined from knot conformations into non-negative real numbers. The infimum of an energy function is an invariant which defines (not necessarily unique) "canonical conformations" of knots in three space. Many (infinite) hierarchies of energy functions for knots in the mathematical and physical science literature have been studied, each energy function with its own (characteristic) set of properties. In this paper we examine energy functions, and classify them as either *basic, strong, charge* or *tight*, depending on the properties of the energies for different knot conformations. Knot invariants derived from these energy functions are expected to be useful.

Key words. Knots, Energy of Knots, Properties of Energy Functions, Thickness.

1. Introduction and definitions. The energy of knots was studied as a means of finding canonical conformations of a given knot [2,6,9,15,18]. There are several energy functions which have received attention, and the number of possibilities for alternative energies is infinite [16,17]. However, a closer examination reveals that many energies are related in the sense that they have the same properties when analyzed as knot invariants. Thus, in order to carry out a systematic study of knot energies, we propose a classification scheme for energy functions of knots based on properties of knot invariants derived from these energy functions.

There are several questions one can ask when considering a functional as a knot energy. For example, is it true that only the unknot is at the global infimum of the energy functional, and is the minimal conformation of the unknot a circle (see definition 1.5)? Are infimums of the non- trivial knots attained by nonsingular conformations? How many distinct knot types will have infimums below a given threshold? From a mathematical point of view, the infimum of an energy function will be a knot invariant where we take the infimum of the energy function over all conformations of a given knot within a certain class (for example, differentiable). At the realization of the infimum we hope that there will be a "canonical conformation" of the knot. The definition of an energy function should include finiteness on all non-singular knots, continuity, scale invariance and invariance under rigid motions, and the possible resolution of knot-types from one another. If in addition, the energy function distinguishes the circle from all other knots, and if it is infinite when the knot embedding is singular and assigns values below a given bound to only finitely many different knot types, then it is

* Department of Mathematics, Kennesaw State College, Kennesaw, GA 30144.
† Department of Mathematics, Western Kentucky University, Bowling Green, KY 42101.
‡ Department of Mathematics and Statistics, York University, North York, Ontario, M3J 1P3, Canada.

a particularly good energy function which we will call *semi-ideal*. Another desirable property, called *tight*, is that the knot energy becomes infinite if one tries to pull the knot "tight" (meaning that there is a knotted arc ball pair where the ball becomes smaller and smaller.) There are knot energies which are infinite for singular knot embeddings, but not tight, as we shall see in the next section. If one uses a knot energy function which is not tight to find a canonical form of a knot by minimizing the knot energy, then it may happen that the knot energy is minimized by creating a tiny knotted ball pair on an almost perfect circle. In the limit all knotted curves may approach the circle under energy minimization. We will call a tight semi-ideal energy function *ideal*.

DEFINITION 1.1. *Let $\alpha(s)$ and $\beta(s)$ be two simple closed curves of the same length parametrized by arc length. If the curves are k-times differentiable, then the k-***distance** *between them is defined as*

$$d^k(\alpha(s),\beta(s)) = \min_{A\in O}\{\max_{0\le s\le L, 0\le t\le L}\sum_{i=0}^{k}|A\alpha^{(i)}(s+t)-\beta^{(i)}(s)|\}, \tag{1.1}$$

where L is the length of $\alpha(s)$ and $\beta(s)$, A is a rigid motion (the composition of a translation and a rotation), O is the group of all rigid motions in $\mathcal{R}^3$ and $\alpha^{(i)}(s)$, $\beta^{(i)}(s)$ are the i-th derivatives of $\alpha(s)$ and $\beta(s)$.

Note that the k-th distance between two curves can be achieved at certain A and t_0, i.e., there exist $A \in O$ and $t_0 \in [0, L)$ such that $d^k(\alpha(s),\beta(s)) = d^k(\hat\alpha(s),\beta(s)) = \max_{0\le s\le L}\sum_{i=0}^{k}|\hat\alpha^{(i)}(s)-\beta^{(i)}(s)|$ where $\hat\alpha(s) = A\alpha(s+t_0)$. We leave it to the reader to verify that d^k defines a metric on the set of all simple closed curves of equal length that are k times differentiable, if we define two such curves α and β to be the same when $d^k(\alpha,\beta) = 0$.

DEFINITION 1.2. *Let $C^k(T)$ be the set of all simple closed curves of unit length that are k times differentiable. Then $f : C^k(T) \longrightarrow \mathcal{R}$ is said to be k-continuous on $C^k(T)$ if $\forall\epsilon > 0$, $\exists\delta > 0$, such that $|f(\alpha) - f(\beta)| < \epsilon$ whenever the k-th distance between α and β is less than δ.*

Remark. If $f : C^k(T) \longrightarrow \mathcal{R}$ is k- continuous, then f is invariant under rigid motion, i.e., $f(A\alpha) = f(\alpha)$ for any $\alpha \in C^k(T)$ and $A \in O$. Also, f will be m- continuous on $C^m(T)$ if $m \ge k$ since $C^m(T) \subset C^k(T)$. On the other hand, one can easily find examples in which f is k-continuous but not m-continuous for some $m < k$.

DEFINITION 1.3. *Let $C^k(T')$ be the set of all simple closed curves (of any length) that are k times differentiable. Let $f : C^k(T') \longrightarrow \mathcal{R}$. Then f is said to be scale invariant if we have $f(\lambda\alpha(s)) = f(\alpha(s))$ for any real number λ and $\alpha(s) \in C^k(T)$.*

DEFINITION 1.4. *Let $f : C^k(T') \longrightarrow \mathcal{R}^+$. Then f is called an energy function on $C^k(T')$ if it is scale invariant and f is continuous when restricted to $C^k(T)$.*

The essential properties of an energy function are given in definition 1.4. Let f be an energy function on $C^k(T')$. Then for a given knot type $\mathcal{K}$, the energy of $\mathcal{K}$ denoted $f(\mathcal{K})$ is defined to be the infimum of f over all knots K of knot type $\mathcal{K}$ in $C^k(T')$.

DEFINITION 1.5. *Let $f : C^k(T') \longrightarrow \mathcal{R}^+$, where $k > 0$, be an energy function. Then*

(a) f is called a basic *energy function if $f(\alpha)$ is an absolute minimum if and only if α is a planar circle of radius $L/(2\pi)$, where L is the length of K;*
(b) f is called a strong *energy function if for any given positive number a, there are only finitely many knot types $\mathcal{K}$ such that $f(\mathcal{K}) \leq a$;*
(c) f is called a charge *energy if f goes to infinity when α approaches a closed curve with self-intersections;*
(d) f is called a tight *energy if $f(\alpha_i)$ goes to infinity for any sequence of embeddings α_i of equal length and of the same knot type K, where α_i contains a knotted arc of knot type K in a ball B_i and the diameter of the ball B_i converges to zero.*
(e) f is called a semi-ideal *energy function if it satisfies conditions (a), (b) and (c).*
(f) f is called an ideal *energy function if it satisfies conditions (a), (b), (c) and (d).*

It is obvious but worthwhile to point out that if f and g are both energy functions and satisfy the condition $f(K) \leq g(K)$ for any K, then if f is strong or charge, then so is g.

2. Some examples. In this section, we give several examples of various energy function of smooth knots without detailed proofs. These are all well known energy functions. A careful reader will find from these examples that the conditions *basic, charge, strong* and *tight* do not contain each other.

Example 1. Let $C^2(T')$ be the set of all twice differentiable simple closed curves and let $K \in C^2(T')$ be parametrized by arc length s. Let $\alpha(s)$ be the parametric equation of K with arc-length or unit speed parametrization. Then $|\alpha'(s)| = 1$, and the curvature of K is $\kappa(s) = |\alpha''(s)|$. The total curvature and total squared curvature are well understood energy functions [11,13]. These may be generalized as follows into a hierarchy of energy functions involving $\kappa(s)$: Let L be the length of K and define

$E_p^f(K) = \frac{1}{L}\int_0^L (L\kappa(s))^p ds$, where p is a constant and L is the total length of K. This defines a continuous function over $C^2(T')$ and we can prove that $E_p^f(K)$ is a basic energy function for $p > 1$ but not a strong energy function, nor a charge energy function. The proof of the case $p = 2$ can be found in [11]. For the other cases of p, it is rather obvious that $E_p^f(K)$ is not charge. The fact that it is basic can be obtained by using Hölder's inequality with the well known result of Milnor [13] that any knot has total curvature at least 4π. A constructive proof can show that all $(2, k)$ torus knots have the same minimal energy 4π hence $E_p^f(K)$ is not strong.

We consider tightness next by outlining a proof of tightness if $p > 1$, and proving that $E_p^f(K)$ is not tight if $p \leq 1$. Let B_i be a sequence of knotted arc ball-pairs of radii $r_i > 0$ where $r_i \to 0$ as $i \to \infty$. Let l_i be the arc in B_i and suppose that it has length $|l_i|$ and that l_i is parametrized by $s \in [0, |l_i|]$. The curvature at a point s on l_i is $\kappa_i(s)$. If the B_i are scaled versions of the same knotted arc ball-pair, then $\kappa_i = \kappa_0 \frac{r_0}{r_i}$. The contribution to the energy from l_i is $\int_0^{|l_i|}[\kappa_i(s)]^p ds = \int_0^{|l_0|}[\kappa_0(s)\frac{r_0}{r_i}]^p r_i ds$. This goes to infinity as $r_i \to 0$ if $p > 1$ but is finite if $p \leq 1$. Thus $E_p^f(K)$ is not tight if $p \leq 1$.

Now let $p > 1$ and let B be a ball that contains a knotted arc C. Let $P = \alpha(s_0)$ be one end point of C and let $Q = \alpha(s_1)$ $(s_1 > s_0)$ be the first point on C such that the angle between $\alpha'(s_0)$ and $\alpha'(s_1)$ becomes 1/2 radian. Such point must exist otherwise C cannot be knotted. This implies that $|\alpha'(s) - \alpha'(s_0)| \leq \sqrt{2(1 - \cos 1/2)} < 1/2$ for all $s_0 \leq s \leq s_1$. We claim that $s_1 - s_0 \leq 4r$ where r is the radius of B. If not, we would have

$$\begin{aligned}
|\alpha(s_1) - \alpha(s_0)| &= \left|\int_{s_0}^{s_1} \alpha'(s)ds\right| \\
&= \left|\int_{s_0}^{s_1} (\alpha'(s) - \alpha'(s_0))ds + (s_1 - s_0)\alpha'(s_0)\right| \\
&\geq (s_1 - s_0) - \int_{s_0}^{s_1} |\alpha'(s) - \alpha'(s_0)|ds \\
&\geq (s_1 - s_0) - (s_1 - s_0)/2 = (s_1 - s_0)/2 > 2r.
\end{aligned}$$

This means that Q is outside of B which is a contradiction. From the condition on the angle between $\alpha'(s_0)$ and $\alpha'(s_1)$ it follows that $\int_{s_0}^{s_1}\kappa(s)ds \geq \frac{1}{2}$. Let $u = (s - s_0)/(s_1 - s_0)$, then we have $\int_{s_0}^{s_1} \kappa(s)ds = (s_1 - s_0)\int_0^1 \kappa_1(u)du$ where $\kappa_1(u) = \kappa(s_0 + (s_1 - s_0)u)$. Thus $\int_0^1 \kappa_1(u)du = \frac{1}{s_1 - s_0}\int_{s_0}^{s_1} \kappa(s)ds \geq \frac{1}{2(s_1 - s_0)}$. Finally, by Hölder's inequality, we have

$$\begin{aligned}\int_{s_0}^{s_1} \kappa^p(s)ds &= (s_1 - s_0)\int_0^1 \kappa_1^p(u)du \\ &\geq (s_1 - s_0)\Big(\int_0^1 \kappa_1(u)du\Big)^p \\ &\geq \frac{1}{2^p(s_1 - s_0)^{p-1}} \geq \frac{1}{2^{3p-2}r^{p-1}}.\end{aligned}$$

This shows that if $r \to 0$, then $E_p^f(K) \to \infty$. So $E_p^f(K)$ is tight if $p > 1$.

Example 2. Let K be any C^1-knot and let $c(K)$ be the usual crossover number of K ($c(K)$ is the minimum number of crossing points over all regular diagrams with knot type K). Then $c(K)$ is scale invariant and continuous. $c(K)$ is a strong energy function, but is not basic, not charge and not tight.

Example 3. Let $K \in C^1(T')$ be arclength parametrized and $\alpha(s)$ be such a parametric equation of K. Let L be the length of K so $\alpha(s)$ is of period L. Define

$$g(s,t) = \frac{t}{|\alpha(s) - \alpha(s+t)|},$$

where $0 < t \leq \frac{L}{2}$ and $|\alpha(s) - \alpha(s+t)|$ is the distance between $\alpha(s)$ and $\alpha(s+t)$. Observe that $g(s,t)$ can be extended continuously to $0 \leq t \leq \frac{L}{2}$ if we define $g(s,0) = 1$ for any s, since $\lim_{t\to 0} g(s,t) = \lim_{t\to 0} \frac{t}{|\alpha(s)-\alpha(s+t)|} = 1$ for any s. Define

$$\mathcal{F}(K) = \sup_{0\leq s\leq L, 0\leq t\leq \frac{L}{2}} g(s,t).$$

Then $\mathcal{F}(K)$ is the same as the Gromov distorsion $D_s(K)$ of K defined as

$$D_s(K) = \sup_{u,v\in[0,1], u\neq v} \frac{D(\alpha(u), \alpha(v))}{|\alpha(u) - \alpha(v)|}$$

in [7,8], where $D(\alpha(u), \alpha(v))$ is the distance between $\alpha(u)$ and $\alpha(v)$ along the curve $\alpha(s)$. $\mathcal{F}(K)$ is scale invariant and continuous since $g(s,t)$ has these properties so $\mathcal{F}(K)$ is an energy function. It is obvious that $\mathcal{F}(K)$ is a charge function. We show below that $\mathcal{F}(K)$ (or the Gromov's distorsion of K) is not tight. $\mathcal{F}(K)$ is basic since $\mathcal{F}(K) = \frac{\pi}{2}$ if K is a circle and $\mathcal{F}(K) > \frac{\pi}{2}$ if K is not a circle (See lemma 3.1). It is known that there exists infinitely many knot types with Gromov's distorsion bounded by a fixed number M [7,16], and so $D_s(K)$ is not a strong energy function. In the next paragraph we show that Gromov's distorsion is neither tight nor strong (the argument that Gromov's distorsion is not strong is an alternative to that of reference [16]).

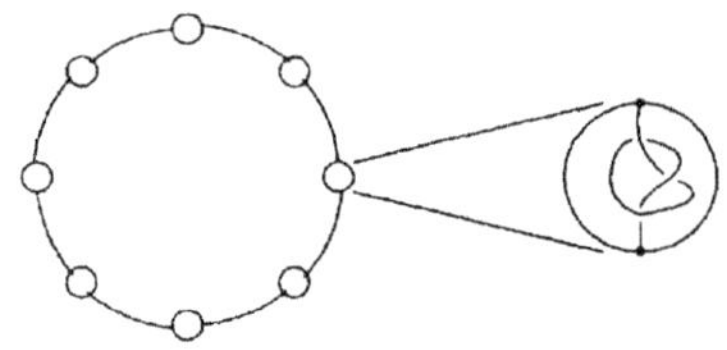

FIG. 1. *A Knot Containing n Copies of Trefoil.*

$\mathcal{F}(K)$ can be defined in a scale invariant manner on any open simple C^1 curve. Thus we can calculate its value on a knotted curve, such as a trefoil. The energy is scale invariant, so the trefoil can be shrunk to be very small without changing its energy. Now let K be a knot with n copies of the trefoil as shown in figure 1 such that each copy is very tiny compare to the length of K such that K takes on the appearance of a circle when viewed macroscopically. Assume furthermore that the copies of trefoils are identical up to rotation and translation. Then the maximum of $g(s,t)$ is attained either in each copy of the small trefoils, or in the macroscopic circle. In both cases we can multiply the copies of the trefoils without bound, and hence $\mathcal{F}(K)$ is not strong. By using only one copy of the trefoils and shrinking it to a point, we see that $\mathcal{F}(K)$ is not tight either. $\mathcal{F}(K)$ can be easily modified into a charge energy function which is not strong nor basic by defining $\mathcal{F}'(K) = \max\{\mathcal{F}(K), k\}$ where $k > 0$ is large enough.

Example 4. In [5,10] the following energy function E^b is defined. Let K be a C^2 simple closed curve in $\mathcal{R}^3$ with a parametric equation $\alpha(s)$. Then

$$E^b(K) = \iint \left(\frac{1}{|\alpha(u)-\alpha(v)|^2} - \frac{1}{D^2(\alpha(u),\alpha(v))}\right)|\alpha'(u)|\,|\alpha'(v)|dudv.$$

Note that this energy function is equivalent to a knot energy function defined by O'Hara [15] which equals $\frac{1}{2}E^b(K) - 2$.

In [5] it is shown that $E^b(K) \geq 4$ where equality holds if and only if K is a circle, so $E^b(K)$ is basic. $E^b(K)$ is a charge energy function by its definition. It is also shown in [5] that $2\pi c(K) + 4 \leq E^b(K)$ where $c(K)$ is the crossover number of the knot K as defined in example 2. So $E^b(K)$ is also a strong energy function. In [16] it is shown that $E^b(K)$ is not tight and hence is semi-ideal.

$E^b(K)$ is a special case in a class of energy functions defined by O'Hara [16,17]. Let K be a C^1 immersion of S^1 in $\mathcal{R}^3$ with unit speed parametrization $\alpha(t)$ (that is, $|\alpha'(t)| = 1$). We define the O'Hara functional of order p and degree j of K by

$$e_j^p(K) = \frac{1}{j}\Big[\int_0^1\int_0^1 \Big(\frac{1}{|\alpha(u)-\alpha(v)|^j} - \frac{1}{D^j(\alpha(u),\alpha(v))}\Big)^p du dv\Big]^{\frac{1}{p}}.$$

Note that $e_2^1(K) = \frac{1}{2}E^b(K) + 2$. It has been shown that $e_j^p(K)$ is strong if $0 < j \leq 2$ and $jp > 2$ [16]. (The case $j = 2$, $p = 1$ is discussed in the above). It is now also known that e_j^p is charge and tight if $jp > 2$ and $0 < j \leq 2$ [16]. On the other hand, if $jp \leq 2$ and $0 < j \leq 2$, then this energy is not tight. It is not known that $e_j^p(K)$ is basic, except in the case discussed above. Finally, $\lim_{j\to 0}\lim_{p\to\infty} e_j^p(K) = \log D_s(K)$ where $D_s(K)$ is the Gromov distorsion defined in example 3 [16].

Example 5. The *c-neighborhood* of a simple closed C^1-curve K is the set of all points with distances less than or equal to c from K. We denote the c-neighborhood of K by K_c. Also, for any $x \in K$, let Σ_x denote the plane that intersects K perpendicularly at x. If c is small enough, then K_c is homotopic to K via strong deformation retract (this is written as $K \sim K_c$) [3]. For any $x \in K$, the component A of $K_c \cap \Sigma_x$ containing x is a meridian disk of K_c which only intersects K at x (for detailed proof see [3]). Saying A is a meridian disk means that A is a topological disk (not necessarily circular) and that $K_c - \text{int}(A)$ is a closed 3-ball with 2 open 2-discs removed from its boundary. A number c with this property is called a *nice* number and the *thickness of* K is defined by

$$t(K) = \sup\{c : \ t \text{ is } nice\ \forall t \in (0, c)\}. \tag{2.1}$$

THEOREM 2.1. *Let K be a C^1 knot of length L, then $\frac{L}{t(K)}$ is not continuous in the C^1 metric but has all the other properties of an ideal energy function.*

Remark. Definitions for different thicknesses of knots are given in [14,12,3]. For these other thicknesses none of these papers establishes continuity in the C^1 metric defined here. In [1] an upper bound of the crossover number of a knot with given thickness (using the definition in [12]) is established.

Proof. Let $E^h(K) = \frac{L}{t(K)}$. For any C^1 knot, we have $t(K) > 0$ hence $E^h(K)$ is defined. It is obvious that $E^h(K)$ is scale invariant hence an energy function.

If K is a circle of radius 1, then the thickness of K is 1 and we have $E^h(K) = 2\pi$. If K is of length 2π but has thickness $c \geq 1$, then let m_ϵ be a meridian of $K_{c-\epsilon}$ (which is a solid torus homotopic to K). Fix a point X on K and let Y be a point on m_ϵ. Let l_Y be a line containing X and Y. As Y moves along m_ϵ, l_Y will meet K at a point Z_ϵ other than X for some Y_ϵ, since m_ϵ and K are linked. For $\epsilon = \frac{1}{n}$, we will then have a sequence of points $(Y_{\frac{1}{n}}, Z_{\frac{1}{n}})$ on $S^1 \times S^1$. This sequence will have a convergent subsequence since $S^1 \times S^1$ is compact. Let (Y_0, Z_0) be the limit point and let $B(Y_0)$ be the ball of radius c centered at Y_0. Clearly, $B(Y_0)$ does not intersect K at an interior point and the line segment XZ_0 contains a diameter of $B(Y_0)$ hence the length of K is greater than or equal to $2\pi c$. The equality is true only when X and Z_0 are both on $\partial B(Y_0)$ and K is the great circle on $B(Y_0)$. Since K is of length 2π, $c = 1$ and K is a circle. This shows that $E^h(K)$ is a basic energy function.

It is obvious that $E^h(K)$ is a tight and charge energy function so we will now prove that $E^h(K)$ is also a strong energy function. Let $a > 0$ be a constant. We need to show that there are only finitely many knot types for a knot K such that $E^h(K) \leq a$. Let K be any knot of unit length such that $E^h(K) \leq a$, so the thickness of K is at least $\frac{1}{a}$. Let $c = \frac{1}{2a}$ and consider K_c. Let $n = \lfloor 2a \rfloor + 1$ and divide K in n pieces of arclength $\frac{1}{\lfloor 2a \rfloor + 1}$. This defines n points on K with a natural order. Connect the consecutive ones with line segments and we obtain a polygonal knot. Obviously, this polygonal knot is of the same knot type as K. Since the number of crossings in any regular projection of this polygonal knot is bounded by $n^2 \leq (2a+1)^2$, the crossover number of K is also bounded by $(2a+1)^2$. So there are only finitely many knot types for K since there are only finitely many knot types with crossover number less than any given constant.

We now show that $t(K)$ is not continuous. Imagine a C^1 curve which has a piece as shown in figure 2. The dotted line represents the plane Σ_x for the point x on the curve. In this configuration the thickness of the curve is determined by the distance d from the dotted line to the minimal point M since in K_d y is on Σ_x. Now a small C^1 deformation that will move the local maxima below Σ_x will increase the thickness drastically. Thus $t(K)$ is discontinuous. □

3. An ideal energy function. In this section, a new energy function is defined and a complete proof will show that it is an ideal energy function. For a discussion of a discretization of this energy to polygonal curves see [4].

Let $K \in C^3(T')$ be arclength parametrized and $\alpha(s)$ be such a parametric equation of K. Let L be the length of K so $\alpha(s)$ is of period L. We define an energy function $E^e(K)$, related to $E^b(K)$, as follows: Let

$$G(s,t) = L^2\left(\frac{1}{|\alpha(s) - \alpha(s+t)|^2} - \frac{1}{t^2}\right),$$

where $0 < t \leq \frac{L}{2}$. Observe that $G(s,t)$ can be extended continuously to $0 \leq t \leq \frac{L}{2}$ if we define $G(s,0) = \frac{L^2}{12}|\alpha''(s)|^2 = \frac{L^2}{12}\kappa^2(s)$ for any s, since

$$\lim_{t\to 0} G(s,t) = -\frac{L^2}{4}|\alpha''(s)|^2 - \frac{L^2}{3}\alpha'(s)\cdot\alpha'''(s)$$

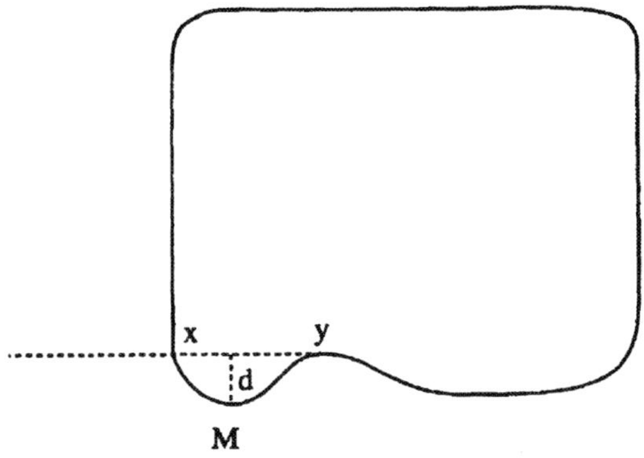

FIG. 2. *$t(K)$ is discontinuous.*

and

$$\alpha'(s) \cdot \alpha'''(s) = -\alpha''(s) \cdot \alpha''(s).$$

Define

$$E^e(K) = \max_{0 \le s \le L, 0 \le t \le \frac{L}{2}} G(s,t).$$

In order to prove that $E^e(K)$ is an ideal energy function, we need the following result:

LEMMA 3.1. *Let $K \in C^1(T')$ be a non circle of length 2π and $\alpha(s)$ be parametric equation of K. Then there exists $s_0 \in [0, 2\pi]$ such that $|\alpha(s_0) - \alpha(s_0 + \pi)| < 2$.*

Proof. If this is not true, then for all s we will have $|\alpha(s) - \alpha(s+\pi)| \ge 2$. Let $t \in [0, 2\pi]$ be such that

$$|\alpha(t) - \alpha(t + \pi)| = \min_{0 \le s \le 2\pi} |\alpha(s) - \alpha(s + \pi)| = r \ge 2.$$

Consider the sphere S centered at the origin O with radius r. The curve $\beta(s) = \alpha(s) - \alpha(s + \pi)$ does not intersect the interior bounded by S. Furthermore, the points $\beta(t)$ and $\beta(t + \pi)$ define a diameter of S since $\beta(t) = -\beta(t + \pi)$. The shortest curve on or outside of S that can join these two points is half of a great circle on the sphere, which is of length πr. Therefore, the length L of the curve $\beta(s)$ is at least $2\pi r$. On the other hand, the length of $\beta(s)$ can be calculated directly as follows.

$$L = \int_0^{2\pi} |\beta'(s)| ds = \int_0^{2\pi} \sqrt{2 - 2\alpha'(s) \cdot \alpha'(s + \pi)} ds. \tag{3.1}$$

Note that in the above we have used the fact that $|\alpha'(s)| = 1$ for any s. Clearly, L is at most 4π since $2 - 2\alpha'(s) \cdot \alpha'(s+\pi)$ is at most 4 for any s. So we have $4\pi \leq 2\pi r \leq L \leq 4\pi$, hence $r = 2$ and $\beta(s)$ is a great circle on S. And this is so if and only if $2 - 2\alpha'(s) \cdot \alpha'(s+\pi) = 4$ for all s. That is, $\alpha'(s) = -\alpha'(s+\pi)$ for all s. But then the point $\frac{\alpha(s)+\alpha(s+\pi)}{2}$ is a constant point and $\alpha(s)$ has constant distance 1 from this point for all s. So $\alpha(s)$ is on a ball with center $\frac{\alpha(s)+\alpha(s+\pi)}{2}$ and with radius 1 with the property $|\alpha(t) - \alpha(t+\pi)| = 2$. The shortest curve joining two diameter points is the half great circle. Since K has length 2π, it has to be a great circle on this ball. This contradicts to our assumption. □

THEOREM 3.2. *$E^e(K)$ is an ideal energy function.*

Proof. $E^e(K)$ is scale invariant and $E^e(K)$ is continuous since $G(s,t)$ is continuous. So $E^e(K)$ is an energy function.

For a unit circle (and hence for any circle), we can easily calculate that

$$G(s,t) = 4\pi^2\left(\frac{1}{4\sin^2\frac{t}{2}} - \frac{1}{t^2}\right) \leq \pi^2 - 4.$$

This function is strictly increasing in t and the maximum $\pi^2 - 4$ is obtained when $t = \pi$. Let K be a non-circle of length 2π and let $\alpha(s)$ be a parametric equation of K. Then by lemma 3.1, there exists a point s_0 such that $|\alpha(s_0) - \alpha(s_0+\pi)| < 2$. Thus, $G(s_0, \pi) = 4\pi^2\left(\frac{1}{|\alpha(s_0)-\alpha(s_0+\pi)|^2} - \frac{1}{\pi^2}\right) > \pi^2 - 4$ hence $E^e(K) > \pi^2 - 4$. This shows that $E^e(K)$ is a basic energy function.

It is obvious that $E^e(K)$ is a charge function but it is not obvious that it is tight. To see this, assume that K has a tiny knotted arc on it. Call a (geometric) ball B on K a minimal knotted arc-ball pair if the ball contains a knotted arc of K (not properly embedded), such that any slightly larger ball with the same origin as B forms a proper knotted arc-ball pair, and no smaller ball can contain a knotted arc of K. We leave it to our reader to prove the existence of a minimum ball pair. Let B be such a minimum ball with a knotted arc C of K in it. It is obvious that $\partial B \cap C$ contains at least three disjoint components where two of them contain the end points of C. Let $A = \alpha(s_0)$ be a point in one component of $\partial B \cap C$ that does not contain the end points of C. Then it is clear that the curvature of K at A is at least $1/r$, i.e., $|\alpha''(s)| \geq 1/r$ where r is the radius of B. This shows that $E^e(K) \geq G(s_0, 0) = (L^2/12)|\alpha''(s_0)|^2 \geq L^2/(12r^2)$. Therefore, $E^e(K) \to \infty$ as $r \to 0$.

Finally, $E^e(K)$ is strong since we have $E^e(K) \geq E^b(K)$ where $E^b(K)$ is defined in Example 4 of last section, which is strong. □

Acknowledgment. The authors thank Joel Langer for pointing out an error in an earlier version of this paper, and Jon Simon for stimulating discussions.

REFERENCES

[1] G. Buck, *Applications of the Projection Energy for Knots*, to appear.
[2] G. Buck and J. Orloff, *Computing canonical conformations for knots*, Top. Appl., **51** (1993), 246–253.
[3] Y. Diao, C. Ernst and Janse Van. Rensburg, *Thicknesses of Knots*, Preprint.
[4] Y. Diao, C. Ernst and Janse Van. Rensburg, *In search of a good polygonal knot energy*, to appear in J. Knot Theo. Ram..
[5] M.H. Freedman, Z-X He and Z. Wang, *Möbius Energy of Knots and Unknots*, Ann. Math., **139** (1994), 1–50.
[6] S. Fukuhara, *Energy of a knot*, in *The Fête of Topology*, Academic Press (1987), 443–451.
[7] M. Gromov, *Homotopical Effects of Dilation*, J. Diff. Geom., **13** (1978), 303–310.
[8] M. Gromov, *Filling Riemannian Manifolds*, J. Diff. Geom., **18** (1983), 1–147.
[9] V. Katritch, J. Bednar, D. Michoud, R.G. Scharein, J. Dubochet and A. Stasiak, *Geometry and physics of knots*, Nature, **384** (1996) 142–145.
[10] R.B. Kusner and J.M. Sullivan, *Möbius Energies for Knots and Links, Surfaces and Submanifolds*, MSRI Preprint, No. 026–94.
[11] J. Langer and D.A. Singer, *Curve Straightening and a Minimax Argument for Closed Elastic Curves*, Topology, **24** (1985), 75–88.
[12] R.A. Litherland, J.K. Simon, O. Durumeric and E. Rawdon, *Thickness of Knots*, Preprint.
[13] J. Milnor, *On the Total Curvature of Knots*, Ann. Math., **52** (1950), 248–257.
[14] Moffatt, *The Energy Spectrum of Knots and Links*, Nature, **347** (1990), 367–369.
[15] J. O'Hara, *Energy of a Knot*, Topology, **30** (1991), 241–247.
[16] J. O'Hara, *Family of Energy Functionals of Knots*, Top. App., **48** (1992), 147–161.
[17] J. O'Hara, *Energy Functionals of Knots II*, Top. App., **56** (1994), 45–61.
[18] J.K. Simon, *Energy Functions for Polygonal Knots*, J. Knot Theo. Ram., **3**, No. **3** (1994), 299–320.

ENERGY AND THICKNESS OF KNOTS*

JONATHAN SIMON†

1. Introduction. Knots in real physical systems, be they rope or DNA loops, have real physical properties; that is a truism. The behavior of *physical knots* depends on the topological types of the knots; that is an experimental observation. Mathematical differences between knot types and, more generally, the whole body of knot theory should help explain the physical behavior; that is a hope. If we make the three simplest knots out of similar "rope" having the same lengths [Figure 1], we see that one kind of knot seems more "tight" or "compact" than another. Somehow this difference will manifest itself in physical systems and should be predictable from the topology.

The goal of connecting topology to physical behavior has been an important motivation for research in recent years on topics such as *self-repelling energy* functions for knots [Fu,O1–O4,BS,BO1–2,KS,Si2,Lo,DEJ], *thickness* of knots [L,Si1,DEJ,KB,LSDR], *edge number* for polygons or lattice knots [R,Me,Jin,Jin2,Ad,JvW,Dia], and probability in random knotting experiments (see fifth paragraph below and other articles in this volume). These all are theoretical, often quite naive, measures of knot complexity.

Despite their simple definitions, which undoubtedly oversimplify most of the detailed physics of a careful elastic rod or other detailed molecular model [He,Di,Ta,Ol,S], simple measures of knot complexity have proven to be good predictors of actual behavior. For example, in the gel electrophoresis of DNA loops, the minimum crossing number of each knot type is a good qualitative predictor of relative velocity [DSKC,DC,DC2,Ketc,SSC,Cr]. And minimum "energy" [Si3] or maximum thickness (= minimum rope-length) [SK] are able to predict in certain experiments which knot type with some minimum crossing number will migrate faster than another with the same minimum crossing number. In fact, there appear to be *quantitative* relations between e.g maximum thickness and relative gel velocity [SK]. The general understanding is that different knot types have different average "sizes" (again, see [Figure 1]), and average size is a basic determinant of gel velocity.

We must be careful to include the caveat *in certain experiments* because one knot type can be faster than another under certain experimental conditions and slower under others [Le] (originally pointed out to the author by D.W. Sumners). The lesson here is that linearly ordering knot types

* The author is supported in part by NSF Grant #DMS9407132.
† Department of Mathematics, University of Iowa, Iowa City, IA 52242.
Email address: jsimonmath.uiowa.edu.

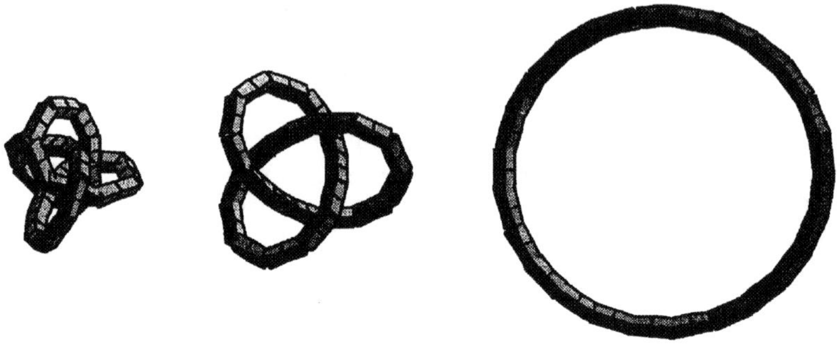

Figure-eight knot *Trefoil knot* *Unknot*

FIG. 1.

according to one or another theoretical measure of complexity is worthwhile but no one ordering can tell the whole story.

Another caveat is that there is more than one notion of "size". If we think of molecules in a gel as tiny bowling balls, then mean radius of gyration may govern velocity. However, for electrophoresis of loops, it may be that the gel also acts like "hooks", catching the loops; so apparent "size" may be a function of the "holes" in a molecule. It is intuitively clear in [Figure 1] that it would be harder to snag the figure-eight knot than the unknot.

Our long-range goal is to understand how the various ideas of knot complexity are related to each other and how they, in turn, relate to physical behavior. In this paper we shall discuss in particular *energy*, *thickness*, and *edge number*, and two of the computer packages that have been developed to study these.

One kind of physical behavior we shall not discuss here is the fascinating topic of *frequency* or *probability* of different knot types in random knotting experiments (laboratory or computer). Interest in the question may be traced to [FW] [Was] or perhaps earlier, while [VLF] may have been the first statistical computer study. The book [MS] and paper [JOS] offer a provocative entré to the subject, [Deg] conjectures a comprehensive model, and [SW1–2,Ry] combine laboratory results with theoretical analysis.

The "story" on relating geometric measures of knot complexity to physical, in particular polymer, behavior is just being written. The discovery that minimum crossing number is a major determinant of (qualitative) relative gel velocity has evolved over 10 years to be a standard assumption, though there is not yet a mechanism based explanation. We showed in [Si3] that minimum knot energy could be a finer discriminant, and in [SK] there is data suggesting even a quantitative relationship between maximum thickness (i.e. reciprocal of energy E_L in Section 3 below) and gel velocity. In Sections 5 and 6 below, we summarize some of the relations that have been established between different energies and other measures of complexity or compaction. The computer simulations of [JW,LJ,KB] are consistent with the theorems and sense of gel behavior we have so far, so perhaps such simulations will suggest future theorems as well. We are working towards being able to justify *and quantify* the relation between minimum energy (or rope-length) and gel velocity by asserting:

(1) The self-repelling knot energies capture a lot of the real (electrostatic and/or bending) self-repelling of polymer loops.
(2) A DNA loop tied in a knot type of low minimum energy will tend to assume conformations that are relatively spread out, with large diameter and large holes, whereas loops tied in knot types of high minimum energy must assume compact conformations.

The author's joint work on energy and on thickness has been with G. Buck, R. Litherland, O. Durumeric, and E. Rawdon. Some preprints are available via anonymous ftp [Si4]. Y.-Q. Wu has been an important colleague, for both mathematical ideas and software development. CS student K. Hunt developed valuable software and data, and J. Martin has explored novel computer algorithms. Many other scientists and fellow mathematicians have, through meetings or correspondence, generously shared their ideas and data.

2. Visualizing and manipulating knots by computer: *ked* and *MING*. A number of computer systems, each with its own strengths and emphases, have been developed to visualize and physically manipulate (virtually, that is) knots, for example [Br,Hu,Sc,Wu,Hg,KB,Rw1-2]; the first four of these are available for downloading via the WWW, as indicated in the References section. These systems were preceded by others whose primary goal (in addition to allowing users to draw and edit knots and links) was combinatorially computing algebraic invariants, including Hoste's *Knotter*, Ochiai's *KnotTheory*, Orloff's *link_edit*, and Litherland's *Grapher*. We describe here the systems *ked* by K. Hunt and *MING* by Y.-Q. Wu.

The programs *ked* and *MING* are fully graphical SGI applications that allow a user to create, manipulate, and energy-minimize knots, represented as polygons. While both programs have all three features, *ked* was developed primarily for careful 3-dimensional manipulation, while *MING* was developed primarily for energy evolution. In [Figure 2] we show the *ked* screen with a 15-segment trefoil knot. In [Figure 3], we illustrate an evolution in *MING* of a trefoil knot; one must remember that the computer is doing the untangling automatically, without the user's intervention.

The starting knot can be built on the screen by editing some initial conformation (via subdividing edges and moving vertices), or read in from a file. The data structure for both programs is simply a number listing how

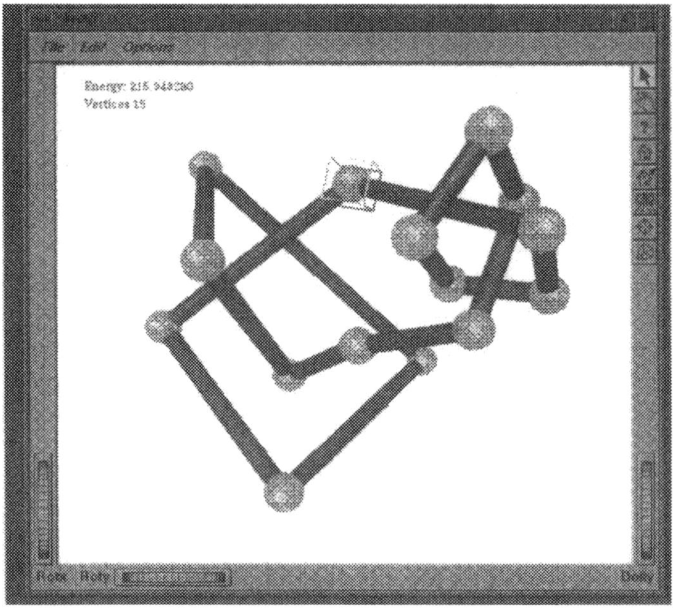

FIG. 2. *The interactive screen for program* **ked**

many vertices there are, followed by subsequent rows each giving the three coordinates x, y, z of the vertices in order. This simplified data structure mitigates against several-component links, though that would be an easy modification which may already have been done by the time this article is appears. (It seems logically impossible to describe a computer system that is constantly being improved.)

Once a knot is determined, the user can manipulate the knot by moving vertices in 3-dimensions. The standard SGI interface makes rotating and translating the whole knot, or zooming, easy. Similarly, the programs offer options for tube thickness, file-handling, etc. An important difference is how they handle the task of moving one vertex independently from the others. In *MING* this is done by click-and-drag in the plane of the screen, using an extra mouse button for the third direction; this is quick, but complicated movements require that the whole knot be rotated so the screen directions become the desired ones. In *ked*, one first selects a vertex to move, then (by rotating a small cube that appears when the vertex is selected, see [Figure 2]) selects a plane in which to move the vertex, then moves in that plane via the usual click-and-drag. Both programs allow the user to subdivide an edge or to remove a vertex and connect the two adjacent vertices with a new edge.

As one moves vertices, it is quite possible that the knot type will change. (But we protect against this in energy evolution). In *ked*, an

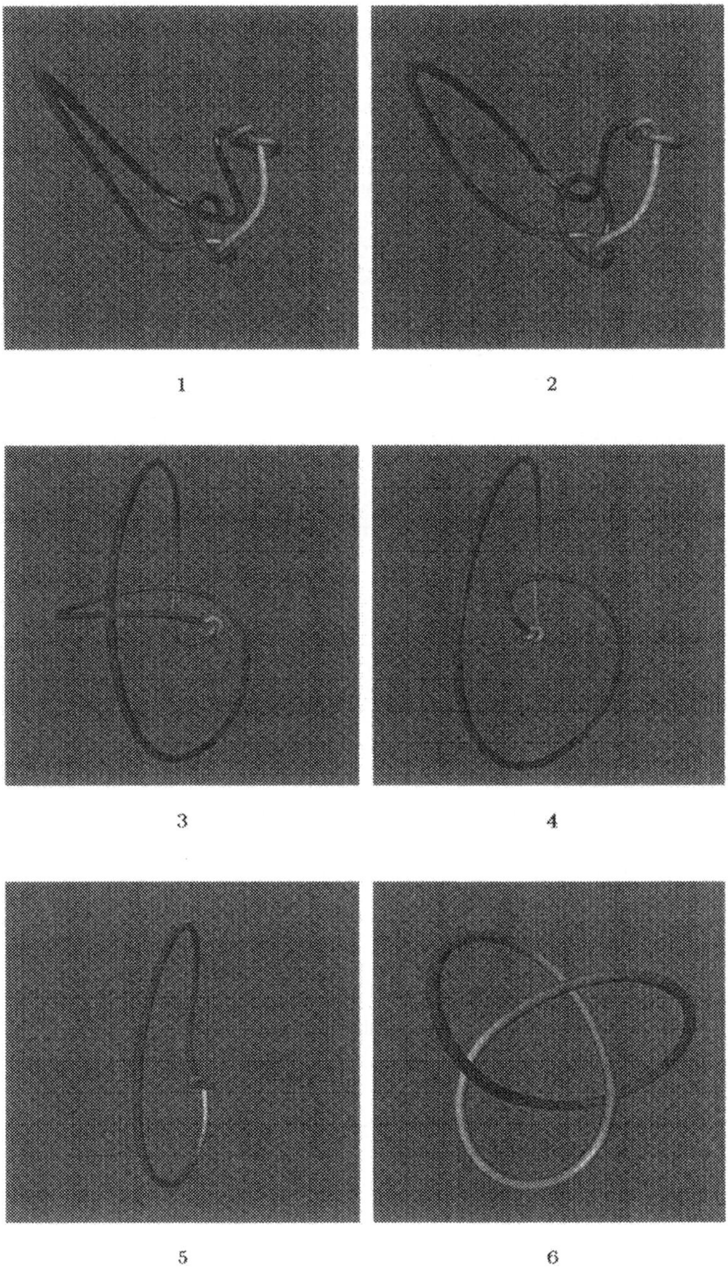

FIG. 3. *Evolution of a knot using* MING

energy number for the knot is constantly displayed; if the user tries to pass an edge of the knot through another, the energy will rise to absurd levels; we have fantasized about having the computer scream or give some other dramatic signal, but this is not yet implemented, so the user needs to keep an eye on the energy number while manipulating the knot. We should also note that neither program tries to keep edge lengths fixed, so the polygons may have some long edges and some short. However, since the energy used would be large if there were large length distortions, energy-minimizing also tends to make edge lengths roughly equal (very roughly - there are examples of apparent minimum energy conformations of knotted polygons for which the edge lengths have a clear pattern, e.g. the 6-segment trefoil knot described in [Si2]). Because of the ability to move polygons precisely, *ked* has been used successfully to find, by hand [Me], presentations of various knots using the smallest possible number of edges. Wu also added to $MING$ a feature of trying to shorten one edge of a knot until it could be deleted (by the computer's decision) without changing knot type, and so automate the process of finding minimum edge-number presentations. This sometimes works well and sometimes is frustrated.

Both programs implement the energy U_{MD} (see [Si2] or the next Section for definition). This energy was specifically developed to study polygonal knots; one can debate whether it is a good energy to use for estimating energies of smooth curves, but it has the advantage of being finite only for nonsingular polygons: self-intersection implies infinite energy. The algorithms used in all automatic movements of the knot in *ked* or $MING$ include controlled step size so the knot-type is preserved.

Both programs offer energy minimization, though just in the form of monotone flow towards a local minimum, not (yet) pursuit of global minima e.g. via simulated annealing. In fact, Wu has found a tangled unknot that apparently is a local minimum; however, the polygon untangles itself (under energy flow) when sufficiently subdivided, so this does not lead, at least in some obvious way, to a nontrivial local minimum for the smooth energies. The energy U_{MD} is a piecewise smooth function on the space of polygons, so at first it seemed pointless to seek a gradient type flow. Program *ked* implemented a simple random perturbation method: shake the knot and move to the new position if the energy is lower than before. Later, a modified random shaking method suggested by J. Martin, which seeks to take advantage of bad guesses, was added as an alternative, so the user can switch between them. Random perturbation methods are slow for knots with many edges, but provide a reassuring computational test that an alleged (local) minimum really is stable. The program $MING$ implements a gradient-based flow. We are aware of the danger of being mislead towards conformations for which the energy is singular but not minimized, though these have not appeared in any of the many trials that have been run. The exact nature of the critical set and critical behavior of U_{MD} remains a subject of study.

3. Definitions of knot energies. For our purposes, a *knot* is a smooth (e.g. class C^2) or polygonal simple closed curve embedded in 3-space $\mathbb{R}^3$.

Topologists have speculated informally about what would happen if a knotted string were somehow given an electric charge and allowed to repel itself. When one tries to make the thought-experiment mathematically precise, the most naive definition of a potential energy function (integral over $K \times K$ of $\frac{1}{|x-y|}$, for a charged knot has mathematically unacceptable properties: because of near-neighbor interactions, it is infinite for all curves, while its sensitivity to distant interactions is too low to prevent strings from passing through each other [BS]. Various investigators have departed from simplest physics to obtain functions that are finite and prevent curves from changing knot type. Typically (but not always, e.g. E_L), the energies are expressed as integrals over the curve (so finite sums for polygons) and, to remove the near-neighbor effect, one either subtracts a term representing the corresponding energy contribution for a standard round circle, or multiplies an integrand by a weight that has the right order of zero for points close together along the curve.

In addition to the functionals discussed below, a number of others have been proposed and studied, such as [BS,O2,DEJ,KS,Lo,Mo].

3.1. Energy for polygonal knots. The *minimum distance energy* for a polygonal knot K is [Si2]: For each pair X, Y of nonconsecutive edges of K, compute the minimum distance between the segments, $MD(X,Y)$ and sum:

$$U_{MD}(K) = \sum_{X,Y \text{ nonconsecutive}} \frac{length(X) \times length(Y)}{[MD(X,Y)]^2} .$$

To consider knots with varying numbers of segments, it is helpful to normalize e.g. by subtracting the energy associated to a standard regular n-gon [Si3]; we denote this regularized polygon energy $\tilde{U}_{MD}(K)$. We have computationally estimated minimum values for $\tilde{U}_{MD}$ for various knot types ("topological ground state energies") [Si3 and subsequent improvements]; the numbers compare well with the values [KS] for E_O, where E_O is the O'Hara/"conformal" energy defined below, and (at least qualitatively) with DNA gel experiments, so we believe that $\tilde{U}_{MD}$ and E_O are measuring analogous, physically meaningful attributes. Although a proof is not yet established, computer experiments by Y.Q. Wu and J. Sullivan reinforce (and correct an earlier version of) our conjecture that the energy $\tilde{U}_{MD}$ applied to polygonal approximations K_n of a smooth knot K gives $2\tilde{U}_{MD}(K_n) \to U_O(K)$.

3.2. Energies for smooth knots. Let the (smooth) knot K be given as a parametrized curve $x(t)$. Then we denote by x, y arbitrary points $x(t), y(t)$ in K and we write $\dot{x}(t)$ for the derivative. We also use: $\rho_{xy} = |x-y|$ (or just ρ), $\boldsymbol{r}_{xy} = \frac{x-y}{|x-y|}$, and $d\boldsymbol{x} = \dot{x}(t)dt$, a line element.

The simplest attempt at a knot energy would be $E_\alpha(K) = \iint\limits_{K\times K} \frac{1}{\rho^\alpha}$ with $\alpha = 1$. However this integral diverges for $\alpha \geq 1$, and fails to provide an infinite barrier to self-crossing for $\alpha \leq 1$. It is common to take $\alpha = 2$ to obtain scale invariance and the desired barrier to changing knot-type. But the integral then diverges. With these considerations in mind we define four energy functions.

The *normal energy* is (equivalent to [BO1]):

$$E_N(K) = \int\int \frac{|d\boldsymbol{x} \times \boldsymbol{r}|^2}{\rho^2} \quad .$$

We can interpret $d\boldsymbol{x} \times r$ as $sin\alpha$, where α is the angle between the chord $(y-x)$ and the tangent direction $d\boldsymbol{x}$, or as $cos\theta$, where θ is the angle between $(y-x)$ and the normal plane to K at x.
The *symmetric energy* [BS2,3] is:

$$E_S(K) = \int\int \frac{|d\boldsymbol{x} \times \boldsymbol{r}||d\boldsymbol{y} \times \boldsymbol{r}|}{\rho^2} \quad .$$

The *O'Hara (also Möbius or conformal) energy* [O1,FHW,KS] is:

$$E_O(K) = \int\int \frac{1}{\rho^2} - \frac{1}{arclength(x,y)^2} \quad .$$

Another "energy" comes from visualizing a knot as actually made of some rope, with a positive thickness. What is the thickest (or shortest) piece of rope one needs to tie a given knot type? To correct for varying lengths, we normalize K to have its length equal 1 or define R(K) to be a ratio of radius/length. To get an energy satisfying Properties 1-4 below, we use the reciprocal of thickness, which we may call the *rope-length* of K (this natural idea also was suggested by [DEJ]). The definition and basic results on thickness were announced in [L,Si1]. Improved theorems and detailed proofs are in [LSDR]. We have now been able [BS2,BS3] to relate rope-length to other energies and so to crossing number. Thus we are establishing a mathematical foundation for computer studies such as [KB,Rw1-2].

Let N_x denote the normal plane to K at x, and $N(x,R)$ the solid disk of radius R centered at x contained in N_x. How large can the radius of such disks be before they run into each other? That is our model of "rope" thickness. Define $R(K) = sup\{\, R : N(x,R) \cap N(y,R) = \emptyset \;\; \forall x \neq y \in K \,\}$.

The *rope-length* or *length-energy* of K is defined as:

$$E_L(K) = \frac{arclength(K)}{R(K)} \quad .$$

This R(K) is equivalent to the radius of injectivity of the projection map $e : N(K;\mathbb{R}^3) \to \mathbb{R}^3$ around the 0-section of the normal bundle. We have a characterization of the thickness of a curve in terms of its curvature and critical self-distances; this is given in Section 4 below.

Each of these functionals, as well as the energy for polygons U_{MD}, satisfies the following properties (among others of course):

1. $E(K)$ is invariant under Euclidean isometry and change of scale;
2. $0 < E(K) < \infty$ if K is not self-intersecting;
3. For knots of fixed total length (or lengths uniformly bounded away from 0), if parts of a knot are brought together to make the knot self-intersect, the energy tends to $+\infty$.
4. There are only finitely many knot types realized below any given energy level, and below some level, only unknots.

Property 1 says E is determined by the shape of a knot, not the size. Properties 2 and 3 tell us that E separates knot-types with infinitely high "potential energy walls", important for trying to prove existence of minimum energy conformations for each knot type; we like to think of these as *canonical conformations*. Property 4 says that energy provides a reasonable taxonomy for knots. One way to establish Property 4 is to prove that each of these energies bounds (2π or 4π times) the crossing number [BO1,BS2,BS3,FHW,Si2]. This is discussed further in Section 5 below.

3.3. Average crossing number. The crossing number of a knot is not an "energy" but it is one of the most fundamental geometric complexity measures.

When a knot in 3-space is projected into a plane, for almost all choices of direction, the projected curve is immersed and one can count the number of self-crossings. The minimum over all projection directions is the familiar *crossing number* $\mathrm{cr}(K)$. Alternatively, one can count the crossings for all the generic directions, and average to produce the *average crossing number* $acn(K)$. A very useful theorem of [FHW] tells us that $acn(K)$ can be computed as a double integral over $K \times K$, which facilitates comparison with energies. We have $acn(K) = \frac{1}{4\pi}\iint\limits_{K\times K} \frac{\|[d\boldsymbol{x}, d\boldsymbol{y}, \boldsymbol{r}]\|}{\rho^2}$, where the numerator of the integrand is the positive triple scalar product of the three vectors. The integral is a modification of Gauss's formula for the linking number of two space curves.

4. Theorems about thickness of knots. We present here a summary of the results on thickness (hence on E_L) in [LSDR] and some additional observations. The summary of [LSDR] is almost the same as given in [BS2]. In addition to their own interest, these results are used to establish various inequalities relating E_L to other energies and to $acn(K)$.

For this section, normalize the smooth knot K to have total length $= 1$. Thus the rope-length E_L is just $1/R$, where R is the actual radius thickness of K.

THEOREM T1. *Maximum curvature of* $K \leq R(K)$.

There actually is a complete characterization of $R(K)$. Define a pair of points (x, y) of K to be *critical* if the chord vector $(x - y)$ is perpendicular to the tangent at x or perpendicular at y. The minimum of all distances $|x-y|$ for such critical pairs is called the *self-distance* of K [Ku]. Let $R_2(K)$ be half of this self-distance. Then we have:

THEOREM T2. *$R(K)$ equals the minimum of the minimum radius of curvature of K and $R_2(K)$. In particular, $R(K)$ is bounded by each of these.*

REMARK. It is clear that $R(K)$ cannot exceed half the *doubly critical self-distance*, that is the minimum distance between pairs of points where the chord is perpendicular to both tangents; however [LSDR] goes on to establish the result for half the *singly critical* self-distance. This may be a bit surprising, as the singly-critical self-distance of a curve is, in general, smaller than the doubly-critical self-distance (e.g. for any ellipse that isn't a circle). But when self-distance is "in control" of thickness, then the two minima coincide. In particular, the thickness $R(K)$ is at most half the singly-critical self-distance of K.

There also is a relation between thickness and the ratios of chord length to arclength along K. For points x, y on K, let $arc(x, y)$ denote the minimum arclength along K between the two points.

THEOREM T3. *Suppose K is a (C^2) knot of total length 1 and thickness R. Then for any x, y on K,*

$$\frac{arc(x,y)}{|x-y|} \leq \frac{1}{4R} .$$

REMARK. In terms of the rope-length $E_L(K)$, Theorem T3 says

$$\frac{arc(x,y)}{|x-y|} \leq \frac{1}{4} E_L .$$

REMARK. The supremum of this ratio over all pairs of points of K is called the *distortion* of K [G,O2].

The following is how one shows that the energy $E_L(K)$ satisfies Property 4 above. Specific formulas for the bounds in this theorem are given in the next Section.

THEOREM T4. *Given a lower bound on $R(K)$ (i.e. an upper bound on $E_L(K)$), one can deduce an upper bound on the bridge number of K (from curvature); the number of sticks needed to represent the knot type $[K]$ as a polygon (from a close analysis of local behavior); and the minimum crossing number of the knot type $[K]$ (from stick number).*

4.1. Additional observations.

4.1.1. Different definitions of thickness. The version of thickness used in [KB] is based directly on polygons, not overtly an approximation of a thickness for smooth curves. In fact [Rw2], if approximating polygons K_n are inscribed in a smooth curve K, then the thickness of K_n as defined in [KB] converges to the minimum of $\frac{1}{2}$ the minimum radius of curvature and $R_2(K)$ (compare Theorem T2 above). This is why the thickness of a standard round circle of radius 1 would be 1 according to [LSDR] but $\frac{1}{2}$ according to [KB]. In [DEJ] there is a suggestion of a definition of thickness that might extend to non-smooth curves.

4.1.2. Existence of thickest knots. For the C^2 thickness defined here and studied in [LSDR], thickness is C^2 continuous ([L2] and different proofs by Durumeric and by Rawdon). Existence of "thickest" knots for each knot type is more problematic; there are results (Litherland, Durumeric) saying that they exist if one does not insist that they are C^2 smooth (so this requires extending the definition to C^1 curves), and results suggesting that optima in general won't be C^2. Thus, in simulations such as [KB] or [Rw1-2], the computer shows convergence of of polygons (with a fixed number of edges) to one that maximizes a polygonal thickness function, but there is no guarantee that there exists some optimal smooth curve that the polygons approximate.

5. Relations between knot energies and crossing number. Ropelength is the "strongest" knot energy as it bounds each of the others (and the implications cannot be reversed) as well as edge-number (edge$[K]$ = fewest edges for a polygon of that knot type) , curvature, Kuiper's *self-distance* and Gromov's *distortion* as discussed above.

The relations established between $E_L(K)$ and crossing number are of two kinds, involving either the average crossing number (hence the min) or directly the minimum for the knot type. The first line of inequalities listed below is established in [BS2]; the others come from [FHW], [LSDR] and [R] respectively.

$$11\, E_L(K)^{4/3} \geq E_N(K) \geq E_S(K) \geq 4\pi \;\; acn(K)$$

$$E_O(K) \geq 4\pi \;\; acn(K)$$

$$\text{edge}[K] \leq 1 + \frac{E_L[K]}{\pi} \qquad \text{and}$$

$$\text{cr}[K] \leq \frac{(\text{edge}[K]-1)(\text{edge}[K]-4)}{2} \quad .$$

Note that these inequalities give us two kinds of relation between $E_L(K)$ and crossing number. If we bound crossings from edges, and edges from thickness, we obtain a bound on the minimum crossing number of the knot type of order ${E_L}^2$. But we also have a bound on average crossing number of the given curve (so it also is a bound on minimum for the knot-type) and it is of lower order, hence better for very complicated knots. For relatively simple knots, the quadratic bound is lower than the (4/3) power bound because the coefficients are smaller.

Just to make this very concrete, suppose we are given 1 meter of rope that has a diameter of 2 cm. So the ratio of length to radius is 100. Which knots can we make, that is, which knots K have $E_L(K) \leq 100$? From the (4/3) power bound, we have (approximately) $acn(K) \leq 406$. Can we actually make a knot that comes close to being this tangled? Continuing, the edge-number bound tells us that edge$[K] \leq 32$ and, from that, the minimum crossing number of the knot type of K is at most 434 (so the $\frac{4}{3}$ power bound is lower than the quadratic bound obtained from edge number). If the length of 2 cm diameter rope is less than 15 cm (more precisely $< 5\pi$) then the edge number of any knot that can be made is at most 5, hence an unknot. The fourth inequality says $[K]$ has minimum crossing number ≤ 2, a sharp estimate, and the first inequality says approximately $acn(K) \leq 32$, a much worse estimate. So perhaps the various inequalities can be improved. The theorem, that it takes a length-to-radius ratio over 5π to tie a nontrivial knot, seems off by a factor of two, in the sense that hand experiments with rope and various computer experiments as well all agree that one needs about twice that length.

We originally hoped for some linear relationship between E_L and E_N (or other inverse square energies); this remains a topic of study. The data in [KB] suggest that there may exist a linear relation between $E_L(K)$ and $acn(K)$ for a particular family of conformations, namely those that minimize E_L; what we know for sure is that each lies within a nonlinear cone around the other for all conformations.

It is an interesting exercise to tabulate some of the estimates for minimum ropelengths of various knots (for closed loops [B,Rw1-2,St], and open rope [DEJ]), along with various energy minima, and see how they compare. There are two general observations: First, all seem to move roughly together, producing (usually) the same ranking of knots; second, as the knots get more complicated, there is evident bunching, where margins of experimental error may overwhelm the perceived differences between one knot type and another. Relationships may look approximately linear if one examines the first few knots, soften to curves for several more, and then wiggle without evident pattern as the knots go on. Similar observations are evident if one generates a few "random" knots and computes various of the energies for these. Our current belief is that there may be regular relationships only for minimum energy or maximum thickness conformations such as suggested in [KB], but that, in general, relationships will involve inequalities and (nonlinear) cones rather than equalities and simple graphs.

J. Cantarella, D. DeTurck, and H. Gluck have also obtained a (4/3) power bound on crossing number in terms of thickness.

As an additional methodological note, G. Buck initially estimated E_L for several prime and composite knots by finding actual physical conformations of rope (see also [DEJ]). These calculations have been in excellent agreement with those done by computer for lower crossing number knots [KB,Rw1-2], and have allowed us to estimate E_L for knots that as yet cannot be handled by computer. The rope also seems adept at discovering several local minima for a given knot, and Rawdon's computer work suggests it is endemic there as well. Tangible physical knots remain a good check on our theories and computations.

6. Relations between energy and compaction. Tightly packed knots should have high energy. If it is true that the knot energies capture a lot of the actual tendency of molecules to self-repel, then the chemical knots will tend to have time averaged conformations near as low energy as the topology permits; thus complicated knots, as measured by minimum energy for the knot type, will be on average more compact (hence faster in gel electrophoresis) than simpler knots.

The numbered theorems in this section are taken from [BS2], and we omit the proofs here. Similar theorems should exist for all the energies.

THEOREM C1. *Let K be an arclength L knot contained in a ball of radius D. Then*

$$E_N(K) \geq \frac{L}{4D} .$$

Proof. This requires a careful geometric argument. □

LEMMA C2.1. *Let K be an arclength L knot such that the entire (singular) tube of radius $R(K)$ is contained in a ball of radius D. Then*

$$E_L(K) \geq (\sqrt{\frac{3}{4}})(\frac{L}{D})^{\frac{3}{2}} .$$

Proof. This follows from Pappus' theorem for curves in 3-space. □

THEOREM C2. *Let K be an arclength L knot contained in a ball of radius D. Then*

$$E_L(K) \geq (\sqrt{\frac{3}{32}})(\frac{L}{D})^{\frac{3}{2}} .$$

Proof. If all of K is contained in the ball the, in particular, a pair of points x, y having maximum distance has $|x - y| \leq 2D$. For such points, the chord vector $(x - y)$ is perpendicular to each tangent (we only need one), so by Theorem T2, $R(K) \leq D$. Since each point of the singular tube about K of radius $R(K)$ has distance $\leq R(K)$ to some point of K, we conclude that the whole singular tube is contained in a ball of radius $2D$. The theorem then follows from Lemma C2.1. □

REFERENCES

[Ad] C. Adams. B. Brennan, D. Greilsheimer and A. Woo, *Stick numbers and compositions of knots and links*, J. Knot Theory and its Ramif., to appear.

[B] G. Buck, *On the energy and length of a knot*, Talk in Special Session on Physical Knot Theory, Amer. Math. Soc., Iowa City, March 1996.

[B3] G. Buck, *Random knots and energy: elementary consideration*, J. Knot Theory and its Ramif. **3** (1994) (reprinted in book Random Knotting and Linking, K.C. Millett and D.W. Sumners (eds), World Scientific, 1994).

[BO1] G. Buck and J. Orloff, *A simple energy function for knots*, Topology and its Appl. **61** (1995), 205–214.

[BO2] G. Buck and J. Orloff, *Computing canonical conformations of knots*, Topology and its Appl. **51** (1993), 246–253.

[Br] K. Brakke (further devel. by J. Sullivan), *SURFACE EVOLVER*, Geometry Center University of Minnesota, http://www.geom.umn.edu/software, Program for visualizing and energy minimizing knots.

[BS] G. Buck and J. Simon, *Knots as dynamical systems*, Topology and its Appl. **51** (1993), 229–246.

[BS2] G. Buck and J. Simon, *Energy and length of knots*, Lectures at Knots '96, Proc. of summer 1996 Intl. Conf., Tokyo, S. Suzuki (ed.), World Scientific Publ. (1997), 219–234.

[BS3] G. Buck and J. Simon, *Thickness and crossing number of knots*, announced in preceding conf., to appear, Topology and its Applications.

[Cr] N.J. Crisona, R. Kanaar, T.N. Gonzales, E.L. Zechiedrich, E.L. Klippel, and N.R. Cozzarelli, *Processive recombination by wild-type Gin and an enhancer-independent mutant. Insight into the mechanisms of recombination and strand exchange*, J. Mol. Biol. **243** (1994), 437–457.

[Dia] Y. Diao, *Minimal knotted polygons on the cubic lattice*, J. Knot Theory and its Ramif. **2** (1993), 413–425.

[DC] P. Dröge and N.R. Cozzarelli, *Recombination of knotted substrates by Tn3 resolvent*, Proc. Nat. Acad. Sci. U.S.A. **86** (1989), 6062–6066.

[DC2] P. Dröge and N.R. Cozzarelli, *Topological structure of DNA knots and catenanes*, Methods in Enzymology **212** (1992), 120–130.

[De] F.B. Dean, A. Stasiak, T. Koller, and N.R. Cozzarelli, *Duplex DNA knots produced by escherichia colo topoisomerase I*, J. Biol. Chem. **260** (1985), 4975–4983.

[Deg] T. Deguchi and K. Tsurusaki, *A statistical study of random knotting using the Vassiliev invariants*, J. Knot Theory and its Ramif. **3** (1994), 321–353 (reprinted in book Random Knotting and Linking, K.C. Millett and D.W. Sumners (eds), World Scientific, 1994).

[DEJ] Y. Diao, K. Ernst, and E.J.J. VanRensburg, *Energies of knots* (preprint, 3/95); *Knot energies by ropes* (preprint 4/96).

[Di] D. Dichmann, Y. Li, and J.H. Maddocks, *Hamiltonian formulations and symmetries in rod mechanics*, Mathematical Approach to Biomolecular Structure and Dynamics (J.P. Mesirov, K. Schulten, and D.W. Sumners, ed.), Proc. of 1994 IMA Summer Program on Molecular Biology, IMA Volume no. 82, Springer-Verlag, 1996, 71–113.

[DPS] Y. Diao, N. Pippenger, and D.W. Sumners, *On random knots*, J. Knot Theory and its Ramif. **3** (1994) (reprinted in book Random Knotting and Linking, K.C. Millett and D.W. Sumners (eds), World Scientific, 1994).

[DSKC] F.B. Dean, A. Stasiak, T. Koller, and N.R. Cozzarelli, *Duplex DNA knots produced by escherichia coli topoisomerase I*, J. Biological Chemistry **260** (1985), 4975–4983.

[FHW] M. FREEDMAN, Z.-X. HE, AND Z. WANG, *Möbius energy of knots and unknots*, Annals of Math. **139** (1994), 1–50.

[FW] H.L. FRISCH AND E. WASSERMAN, *Chemical topology*, J. Am. Chem. Soc. **83** (1961), 3789–3795.

[Fu] S. FUKUHARA, *Energy of a knot*, A Fête of Topology: Papers Dedicated to Itiro Tamura (Y.T. Matsumoto and S. Morita, ed.), Academic Press, New York, 1988, 443–451.

[G] M. GROMOV, *Filling Riemannian manifolds*, J. Diff. Geom. **18** (1983), 1–147 (see p. 113).

[He] J.E. HEARST AND Y. SHI, *The elastic rod provides a model for DNA and its functions*, Mathematical Approach to Biomolecular Structure and Dynamics (J.P. Mesirov, K. Schulten, and D.W. Sumners, ed.), Proc. of 1994 IMA Summer Program on Molecular Biology, IMA Volume no. 82, Springer-Verlag, 1996, 59–70.

[Hg] M. HUANG, Univ. Illinois-Chicago, http://www.eecs.uic.edu/~mhuang/research.html, Program for visualizing and energy minimizing knots.

[Hu] K. HUNT, *KED*, University of Iowa, http://www.cs.uiowa.edu/~hunt/, Program for visualizing, manipulating, and energy minimizing polygonal knots.

[JvW] E.J. JANSE VAN RENSBURG, *Lattice invariants for knots*, Mathematical Approach to Biomolecular Structure and Dynamics (J.P. Mesirov, K. Schulten, and D.W. Sumners, ed.), Proc. of 1994 IMA Summer Program on Molecular Biology, IMA Volume no. 82, Springer-Verlag, 1996, 11–20.

[JOS] E.J. JANSE VAN RENSBURG, E. ORLANDINI, D.W. SUMNERS, M.C. TESI AND S. WHITTINGTON, *Topology and geometry of biopolymers*, Mathematical Approach to Biomolecular Structure and Dynamics (J.P. Mesirov, K. Schulten, and D.W. Sumners, ed.), Proc. of 1994 IMA Summer Program on Molecular Biology, IMA Volume no. 82, Springer-Verlag, 1996, 21–37.

[JW] E.J. JANSE VAN RENSBURG AND S. WHITTINGTON, *The dimensions of knotted polygons*, J. Phys. A: Math. Gen. **24** (1991), 3935–3948.

[JW2] E.J. JANSE VAN RENSBURG AND S. WHITTINGTON, *The knot probability in lattice polygons*, J. Phys. A: Math. Gen **23** (1990), 3573–3590.

[JW3] E.J. JANSE VAN RENSBURG AND S. WHITTINGTON, *The BFACF algorithm and knotted polygons*, J. Phys. A: Math. Gen. **24** (1991), 5553–5567.

[Jin] G.T. JIN AND H.S. KIM , *Polygonal knots*, Jour. Korean Math. Soc. **30** (1993), 371–383.

[Jin2] G.T. JIN, *Polygon indices and superbridge indices of torus knots and links*, preprint 3/96.

[Ketc] R. KANAAR, A. KLIPPEL, E. SHEKHTMAN, J.M. DUNGAN, R. KAHMANN AND N.R. COZZARELLI, *Processive recombination by the phage Mu Gin system: Implications for the mechanisms of DNA strand exchange, DNA site alignment, and enhancer action*, Cell **62** (1990), 353–366.

[KB] V. KATRITCH, J. BEDNAR, D. MICHOUD, R. SCHAREIN, J. DUBOCHET AND A. STASIAK, *Geometry and physics of knots*, Nature **384**, Nov. 1996, 142–144.

[KS] R. KUSNER AND J. SULLIVAN, *Möbius energies for knots and links, surfaces and manifolds*, Geometry Center Research Report GCG64 (1993 rev. 1994), University of Minnesota, to appear in Proc. of 1993 Georgia Top. Conf..

[Ku] N. KUIPER, pers. corresp. with J. O'Hara, see [O2] *loc cit.*

[L] R. LITHERLAND, *Thickness of knots*, Talk in Workshop on 3-Manifolds, Univ. Tennessee, 1992.

[L2] R. LITHERLAND*, J. SIMON, AND O. DURUMERIC, *Thickness of knots*, Talk in AMS Special Session on Physical Knot Theory, Iowa City, 3/96.

[LCJ] H.A. Lim, M.T. Carroll, and E.J. Janse van Rensburg, *Electrophoresis of knotted DNA in a regular and a random electrophoretic medium*, Biomedical Modeling and Simulation (J. Eisenfeld, D.S. Levine and M. Witten, ed.), Elsevier Science Pub., 1992, 213–223.

[Le] S.D. Levene and H. Tsen, *Analysis of DNA knots and catenanes by agarose-gel electrophoresis*, Protocols in DNA Topology and Topoisomerases, vol. I (M. Bjornsti and N. Osheroff, ed.), Humana Press, 1996.

[LJ] H.A. Lim and E.J. Janse van Rensburg, *A numerical simulation of electrophoresis of knotted DNA*, Supercomputer Computations Research Inst. Report **FSU-SCRI-91-163** (1991).

[Lo] S. Lomonaco, *The modern legacies of Thompson's atomic vortex theory in classical electrodynamics*, The Interface of Knots and Physics (L. Kauffman, ed.), (AMS Short Course, Jan. 1995), American Mathematical Society, 1996, 145–166.

[LSDR] R. Litherland, J. Simon, O. Durumeric, and E. Rawdon, *Thickness of knots* (based on [L,Si1]), to appear in Topology and its Applic.

[Me] M. Meissen, *Polygon knot table*, University of Iowa, http://www.math.uiowa.edu/~meissen/PLKnotTable.html.

[MS] K.C. Millett and D.W. Sumners (eds), *Random Knotting and Linking*, World Scientific, 1994.

[Mo] H.K. Moffatt, *The energy spectrum of knots and links*, Nature **347**, Sept. 1990, 367–369.

[O1] J. O'Hara, *Energy of a knot*, Topology **30** (1991), 241–247.

[O2] J. O'Hara, *Family of energy functionals of knots*, Topology Appl. **48** (1992), 147–161.

[O3] J. O'Hara, *Energy functionals of knots*, Topology-Hawaii (K. H. Doverman), (Proc. of 1991 Conference), World Scientific, 1992, 201–214 (Computer program by K. Ahara).

[O4] J. O'Hara, *Energy functionals of knots II*, Topology and its Appl. **56** (1994), 45–61.

[Ol] W. Olson, T. Westcott, J. Martino, G-H Liu, *Computational studies of spatially constrained DNA*, Mathematical Approach to Biomolecular Structure and Dynamics (J.P. Mesirov, K. Schulten, and D.W. Sumners, ed.), Proc. of 1994 IMA Summer Program on Molecular Biology, IMA Volume no. 82, Springer-Verlag, 1996, 195–217.

[Pi] N. Pippenger, *Knots in random walks*, Disc. Appl. Math. **25** (1989), 273–278.

[R] R. Randell, *An elementary invariant of knots*, J. Knot Theory and its Ramif. **3** (1994), 279–286 (reprinted in book Random Knotting and Linking, K.C. Millett and D.W. Sumners (eds), World Scientific, 1994).

[Rw1] E. Rawdon, *The thickness page*, University of Iowa, http://www.math.uiowa.edu/rawdon/thick.html, Preliminary data on program for finding thickest knots.

[Rw2] E. Rawdon, *Thickness of Polygonal Knots*, Ph.D. Thesis, University of Iowa, August 1997.

[Ry] B.Y. Rybenkov, N.R. Cozzarelli, and A.V. Vologodskii, *Probability of DNA knotting and the effective diameter of the DNA double helix*, Proc. Nat. Acad. Sci. **90** (1993), 5307–5311.

[S] T. Schlick, *Pursuing Laplace's vision on modern computers*, Mathematical Approach to Biomolecular Structure and Dynamics (J.P. Mesirov, K. Schulten, and D.W. Sumners, ed.), Proc. of 1994 IMA Summer Program on Molecular Biology, IMA Volume no. 82, Springer-Verlag, 1996, 219–247.

[Sc] R. Scharein, *Knot-Plot*, Univ. British Columbia, http://www.cs.ubc.ca/spider/scharein/, Program for drawing, visualizing, manipulating, and energy minimizing knots.

[Si1] J. SIMON, *Thickness of Knots*, Talk in Special Session on Knotting Phenomena in the Natural Sciences, American Mathematical Society, Santa Barbara, Nov. 1991; *Lecture Notes on Physical Knot Theory* (notes by H. Naka), 1993, from summer 1991 course, Kwansei Gakuin University.

[Si2] J. SIMON, *Energy functions for polygonal knots*, J. Knot Theory and its Ramif. **3** (1994), 299–320 (reprinted in book Random Knotting and Linking, K.C. Millett and D.W. Sumners (eds), World Scientific, 1994).

[Si3] J. SIMON, *Energy functions for knots: beginning to predict physical behavior*, Mathematical Approach to Biomolecular Structure and Dynamics (J.P. Mesirov, K. Schulten, and D.W. Sumners, ed.), Proc. of 1994 IMA Summer Program on Molecular Biology, IMA Volume no. 82, Springer-Verlag, 1996, 39–58.

[Si4] J. SIMON, papers available by anonymous ftp, http://www.math.uiowa.edu/~jsimon/README.html.

[SSC] S. SPENGLER, A. STASIAK, AND N.R. COZZARELLI, *The stereostructure of knots and catenanes produced by phage λ integrative recombination: implications for mechanism and DNA structure*, Cell **42** (1985), 325–334.

[SK] A. STASIAK, V. KATRITCH, J. BEDNAR, D. MICHOUD, AND J. DUBOCHET, *Electrophoretic mobility of DNA knots*, Nature **384**, Nov. 1996, 122.

[St] A. STASIAK, *Ideal forms of knots*, Talk in Special Session on Physical Knot Theory, Amer. Math. Soc., Iowa City, March 1996 (Revised paper to appear Dec. 96 Nature).

[SuW] D.W. SUMNERS AND S. WHITTINGTON, *Knots in self-avoiding walks*, J. Phys. A: Math. Gen. **21** (1988), 1689–1694.

[SW] S.Y. SHAW AND J.C. WANG, *Knotted DNA rings: probability of formation and resolution of the two chiral trefoils*, Science **260** (1993), 533–536.

[SW2] S.Y. SHAW AND J.C. WANG, *DNA knot formation in aqueous solutions*, J. Knot Theory and its Ramif. **3** (1994), 287–298.

[Ta] M. TABOR AND I. KLAPPER, *Dynamics of twist and writhe and the modeling of bacterial fibers*, Mathematical Approach to Biomolecular Structure and Dynamics (J.P. Mesirov, K. Schulten, and D.W. Sumners, ed.), Proc. of 1994 IMA Summer Program on Molecular Biology, IMA Volume no. 82, Springer-Verlag, 1996, 139–159.

[VLF] A.V. VOLOGODSKII, A.V. LUKASHIN, M.D. FRANK-KAMENETSKII AND V.V. ANSHELEVICH, *The knot probability in statistical mechanics of polymer chains*, Sov. Phys.-JETP **39** (1974), 1059–1063.

[Was] E. WASSERMAN, *Chemical Topology*, Scientific American **207(5)** (1962), 94–102.

[WC] S. WASSERMAN AND N.R. COZZARELLI, *Supercoiled DNA-directed knotting by T4 topoisomerase*, J. Biol. Chem. **266** (1991), 73–95.

[Wu] Y.-Q. WU, *MING*, University of Iowa, http://www.math.uiowa.edu/~wu/, Program for visualizing, manipulating, and energy minimizing polygonal knots.

ON DISTORTION AND THICKNESS OF KNOTS*

ROBERT B. KUSNER† AND JOHN M. SULLIVAN‡

Introduction. What length of rope (of given diameter) is required to tie a particular knot? Or, to turn the problem around, given an embedded curve, how thick a regular neighborhood of the curve also is embedded? Intuitively, the diameter of the possible rope is bounded by the distance between strands at the closest crossing in the knot. But of course the distance between two points along a curve goes to zero as the points approach each other, so to make the notion precise, we need to exclude some neighborhood of the diagonal.

Various notions of thickness have been proposed recently. For example, [LSDR] defines a thickness by considering the distance function between points on the curve only where it has critical points. But the definition there also involves the minimum radius of curvature of the curve, and thus is unbounded for polygonal curves. Here we introduce and compare two new families of thickness measures. One makes use of Gromov's concept of distortion (see [Gro1] and [Gro2, p. 114] and [GLP, pp. 6–9]); it applies to all rectifiable curves (including polygons). The other generalizes the notion from [LSDR]. Our main result is a basic inequality (Theorem 5.2) between these measures.

The distortion thickness should permit us to prove the existence of thickest curves of prescribed length (or dually, shortest curves of prescribed thickness) in each knot class; such curves are of interest to chemists and biologists modeling polymers and DNA (see, for example, [KBM+]). Moreover, curvature bounds should follow from the optimality, and we conjecture that, for an optimal knot, all reasonable measures of thickness should be equal.

1. Notation and definitions. We will deal throughout with an embedded rectifiable closed curve γ in $\mathbb{R}^n$, of finite length $L(\gamma)$. Thus we may assume that $\gamma\colon \mathbb{R}/L\mathbb{Z} \to \mathbb{R}^n$ is a Lipschitz parameterization by arclength. (Often we will normalize by rescaling so that $L = 2\pi$.) If $p, q \in \gamma$ are two points on the curve, $|p - q|$ will denote the straight-line (chord) distance between them in $\mathbb{R}^n$, and $d(p, q)$ the arclength distance between them along γ. This arclength distance is always measured the shorter way around γ, so that it is never more than $L/2$. Given any point p on γ, there is a unique *opposite point* p^* such that $d(p, p^*) = L/2$.

* The first author was funded in part by NSF grant DMS 94-04278, and both authors enjoyed the hospitality of the IMA during the Spring 1996 term.

† Department of Mathematics, Lederle Graduate Research Center, University of Massachusetts, Amherst, MA 01003-4515.

‡ Department of Mathematics, University of Illinois, Urbana, IL 61801-2975.

DEFINITION 1.1. *The* distortion *between distinct points p and q on the curve γ is*

$$\delta(p,q) := \frac{d(p,q)}{|p-q|} \geq 1.$$

This is a Lipschitz constant for the inverse of the map γ (which itself has Lipschitz constant 1). The distortion of γ is the maximum distortion between any points:

$$\delta(\gamma) := \sup_{p,q} \delta(p,q),$$

where the supremum is over $(\gamma \times \gamma) \smallsetminus \Delta$, the set of all pairs of distinct points on γ. We can also consider a restricted distortion, only considering opposite pairs:

$$\delta_\circ(\gamma) := \sup_{p \in \gamma} \delta(p, p^*).$$

Clearly $\delta_\circ(\gamma) \leq \delta(\gamma)$ since the supremum is over a subset; also, the supremum in $\delta_\circ$ is achieved, though that for δ may not be.

DEFINITION 1.2. *Given any real number $b \geq 1$, the* distortion thickness *of a curve $\gamma \subset \mathbb{R}^n$ is*

$$\tau_b(\gamma) := \inf_{\delta(p,q) > b} |p-q|,$$

the minimum distance between points of distortion more than b.

The idea here is that on a smooth curve, nearby points have distortion hardly greater than one, so by taking $b > 1$ we eliminate these nearby pairs before taking the minimum distance.

We will define the remaining notions only for $C^{1,1}$ curves, which have a well-defined, Lipschitz continuous tangent vector field, and a curvature κ almost everywhere. A pair of points (p,q) on the curve is called *symmetric* if the unit tangent vectors T_p and T_q make equal angles with the chord $p-q$, that is, $T_p \cdot (p-q) = T_q \cdot (q-p)$. For instance, any two points on a round circle are symmetric. An ordered pair (p,q) of distinct points is said to be *self-critical* [LSDR] if q is a critical point for the distance from p, that is $T_q \cdot (p-q) = 0$. A pair which is symmetric and self-critical is called *doubly self-critical.*

For any $C^{1,1}$ curve γ, define $c_1(\gamma)$ to be the minimum straight-line distance $|p-q|$ among self-critical pairs (p,q), and $c_2(\gamma)$ to be the minimum among doubly self-critical pairs. Also, let $r(\gamma)$ be the minimum radius of curvature along the curve, $r(\gamma) := 1/\sup_\gamma \kappa$; this is positive because a $C^{1,1}$ curve has bounded curvature.

DEFINITION 1.3. *Given a real number $k > 0$, the* curvature thickness *of a curve γ is*

$$\sigma_k(\gamma) := \min\big(kr(\gamma), c_2(\gamma)\big).$$

For $k = 2$, this is (twice) the notion of thickness studied in [LSDR].

Note. If γ is not $C^{1,1}$, its curvature in some sense is unbounded, so we set $\sigma_k(\gamma) = r(\gamma) = 0$.

Since we are interested in minimizing the length of curves with (any one of our notions of) thickness prescribed, or dually, in maximizing thickness for curves of prescribed length, it useful to have a scale-invariant measure for this optimization problem.

DEFINITION 1.4. *The* ropelength *of a curve is its length divided by its thickness. This is the length of a similar curve scaled to have thickness* 1 *(so that it could be made out of a rope of diameter* 1*). Given a topological knot type K, the* ropelength of K *is the infimum ropelength of any curve γ of type K.*

Intuitively, the ropelength of K is the shortest length of rope (of diameter one) which could be used to tie that knot. Of course, there are several notions of ropelength corresponding to our different notions of thickness, but we conjecture they all agree (for reasonable values of k and b), because a curve of shortest ropelength should have nice structure.

Note. Most of the notions we have defined make sense also for any union of immersed rectifiable curves, which need not be closed curves. For a link of several components, we take $d(p, q) = \infty$ if p and q are on different components, so that such pairs are always contenders in the definition of distortion thickness. If the curves are not embedded, there are points p, q such that $d(p, q) > 0$ while $|p - q| = 0$. Such a pair has infinite distortion, and is doubly self-critical, so all our notions agree that such a nonembedded curve has thickness zero.

2. Distortion of curves. In order to understand the definition of distortion thickness τ_b, we need to explore the distortion of curves. In particular, we will see that every closed curve has distortion at least $\pi/2$, so that $b = \pi/2$ is a natural choice.

Recall that the distortion of γ is defined as a supremum of $\delta(p, q) = d(p, q)/|p - q| \geq 1$. We noted before that if γ is not rectifiable or embedded then we will have some $\delta(p, q) = \infty$. The finiteness of $\delta(\gamma)$ is sometimes called a *uniform chord-arc condition* in harmonic analysis [Tor]. We now note that this supremum may be infinite even for an embedded rectifiable curve.

EXAMPLE 2.1. *Consider first a wedge γ of external angle θ. If p and q are at equal distance s from the corner, then $|p - q| = 2s \cos \frac{\theta}{2}$ so*

$\delta(\gamma) = \delta(p,q) = \sec\frac{\theta}{2}$. *Now take a plane curve* γ, *such as* $x^2 = y^3$, *with a sharp cusp at the origin. Then as* p *and* q *approach the origin along* γ *at equal distance* s *from (and on opposing sides of) the cusp, we may approximate a neighborhood of the cusp by narrower and narrower wedges, and conclude that* $\lim \delta(p,q) = \infty$.

However, for a C^1 curve, the distortion $\delta(p,q)$ approaches 1 whenever p and q approach a common point, so we can view δ as a continuous function on the compact space $\gamma\times\gamma$, and it achieves its (finite) supremum somewhere (away from the diagonal).

Note. O'Hara [Oha1] defined a family of energy functionals E_j^p for knots as L^p norms

$$jE_j^p(\gamma) := \left\| |x-y|^{-j} - d(x,y)^{-j} \right\|_p$$

and observed that the distortion is a limiting value $\delta(\gamma) = \exp(E_0^\infty(\gamma))$.

Gromov mentions in [Gro2] that any closed rectifiable curve has distortion at least $\pi/2$. In fact, in [GLP] he proves a somewhat stronger result which includes the following proposition as a consequence. For completeness, we include a proof.

PROPOSITION 2.1. *Given any closed rectifiable curve* $\gamma \subset \mathbb{R}^n$, *the distortion satisfies* $\delta(\gamma) \geq \delta_\circ(\gamma) \geq \pi/2$, *with equality if and only if* γ *is a round circle.*

Proof. Rescale so that $L(\gamma) = 2\pi$, and parameterize the curve by arclength $s \in \mathbb{R}/2\pi\mathbb{Z}$. Opposite points p and p^* have arclength $d(p,p^*) = \pi$, so we want to prove for some p that $|p - p^*| \leq 2$. For then $\delta(p,p^*) \geq \pi/2$ and we are done.

Consider the new curve $f(s) := p - p^* = \gamma(s) - \gamma(s+\pi)$ in $\mathbb{R}^n$. Observe that f is Lipschitz with speed at most two (since by the triangle inequality $|f'(s)| \leq |\gamma'(s)| + |\gamma'(s+\pi)| = 2$ almost everywhere), and that $f(s+\pi) = -f(s)$ for all s.

We want to show that $|f| \leq 2$ somewhere; suppose not, so that f lies *outside* the closed ball of radius two in $\mathbb{R}^n$. Then any arc of f from s to $s+\pi$ is an arc between antipodal points in $\mathbb{R}^n$ which avoids this closed ball, so its length exceeds the distance between antipodal points on a sphere of radius two, which is 2π. But since the parameterization of f has speed at most two, the length of this arc is at most 2π, a contradiction.

Note that the only way to get equality $\delta = \pi/2$ is to have $|f| \equiv 2$, with f tracing out a great circle on this sphere. Then p and p^* must always move in opposite directions, and p also traces out a round circle. □

Even if we had only proved this theorem for smooth curves, the following lemma would extend the result to arbitrary curves; it will also be useful in our discussion of distortion thickness.

LEMMA 2.2. *In the C^0 topology on the space of curves, the distortion, considered as a map to $[1,\infty]$, is lower semicontinuous.*

Proof. In the C^0 topology on parameterized curves (not necessarily with constant speed parameterizations), the position of any point $p = \gamma(t)$ is by definition a continuous function of γ. Thus, for any points p and q the function $|p-q|$ is continuous in the C^0 topology, and the arclength $d(p,q)$ along γ is lower semicontinuous. Hence $\delta(p,q)$ is lower semicontinuous for each pair of points.

Remember that a real-valued function f on any topological space is lower semicontinuous if and only if $f^{-1}\{x : x > y\}$ is open for all $y \in \mathbb{R}$. Thus it is clear that if every member of some family $\{f_\alpha\}$ of functions is lower semicontinuous, then $f := \sup_\alpha f_\alpha$ is also lower semicontinuous. But the distortion $\delta(\gamma)$ is the supremum of $\delta(p,q)$, a family of lower semicontinuous functions, hence is itself lower semicontinuous. □

Note. This lemma, together with the fact that distortion blows up for nonembedded curves, leads us to expect that a curve minimizing the distortion should exist within each knot class. Gromov [Gro1] and O'Hara [Oha1] have independently observed that there are infinitely many knot types with distortion less than some constant $M < 100$. Gromov asked [Gro2, p. 114] if perhaps all knots have distortion under 100. In any case, distortion does not seem to be as useful for distinguishing knots as O'Hara's original energy E_2^1 [Oha,FHW,KS] has been, or as ropelength promises to be.

Note. On a sufficiently smooth curve γ, we can make some observations about any pair (p,q) realizing the maximum distortion. First if we consider the variation of $\delta(p,q)$ due to a perturbation fixing $d(p,q)$, we find that (p,q) must be a symmetric pair, meaning the chord between them makes equal angles θ with the tangents to γ at either end. If p and q are not opposite points on γ, then we can also consider variations increasing or decreasing $d(p,q)$; this shows that $\delta = \sec\theta$. But if $q = p^*$ is opposite p, all we can conclude is that $\delta \leq |\sec\theta|$, which bounds θ near $\pi/2$. Note that a doubly self-critical pair realizing the minimum self-distance $c_2(\gamma)$ cannot also realize the maximum distortion, unless the points are antipodal; some slightly more distant pair will have greater distortion.

3. Distortion and a conjecture in integral geometry. Another approach to proving the proposition from the last section would be to use Crofton's formula and related notions from integral geometry, which relate the average sizes of projected images of a curve to the length of the curve.

Recall that if γ is a plane curve of length L, and we consider the (S^1-worth of) projected images of γ to the line, Crofton's formula says that the average length (with multiplicity) of these projections is $\frac{2}{\pi}L$. Since the multiplicity of the projected closed curve is at least 2 everywhere, we find that the average diameter of the projections is at most L/π. Thus, a plane curve of length 2π has width at most 2 in some direction.

Our proposition follows immediately for plane curves. If γ has length 2π, orient it so that its width horizontally is at most 2. Then find a pair of opposite points p and p^* at the same height (noting that the height difference changes sign as p moves around γ). Clearly $|p - p^*| \leq 2$, while $d(p, p^*) = \pi$, so $\delta_0(\gamma) \geq \pi/2$.

Janse van Rensburg has suggested [JvR] that any space curve could be "unfolded" to a convex plane curve of the same length, while never decreasing the chord distances $|p-q|$. For instance, if a curve touches a supporting hyperplane in two disjoint places, we could reflect one intervening segment to lie below this plane. This procedure certainly never increases the distortion. If we could prove that it converged to a planar curve (perhaps considering only the polygonal case) then we would have a new proof of our proposition, except that it would seem difficult to show circles are the only curves of distortion $\delta = \pi/2$.

The argument sketched above for plane curves could be applied directly to curves in higher dimensions if the following conjecture is true. Again we would find a pair of opposite points on the original curve that have the same height in the direction of the projection, and they would then have distortion at least $\pi/2$.

CONJECTURE 3.1. *If γ is a curve in $\mathbb{R}^n$ of length L, then there is some orthogonal projection to $\mathbb{R}^{n-1}$ in which the image of γ has diameter at most L/π.*

Note. For space curves in $\mathbb{R}^3$ of length L, there are two standard analogues of Crofton's formula. The first deals with projections to a line, and says that the average length (with multiplicities), over the sphere of projections, is $L/2$. Thus some projection to a line has diameter at most $L/4$. The second deals with projections to a plane, and says that the average length of these, over the sphere of possible projections, is $\frac{\pi}{4}L$. Thus some projection to a plane has diameter at most $\frac{\pi}{8}L$, but this is not as good as our conjectured L/π.

4. The distortion thickness. Recall that the distortion thickness of a curve measures the least self-distance among pairs of points with large distortion:

$$\tau_b(\gamma) := \inf_{\delta(p,q)>b} |p-q|.$$

Clearly, as we increase b, $\tau_b(\gamma)$ is nondecreasing, since we take the infimum over smaller sets. For $b \leq 1$, all pairs (p, q) are in contention, except those connected by a straight arc, so $\tau_1 = 0$ for all closed curves. For $b \geq \frac{\pi}{2}$, the infimum may be over the empty set (as it would be for a circle), so the thickness may be infinite. For $b = \frac{\pi}{2}$, although a circle has infinite thickness, no other closed curve does. Thus we concentrate on $b \in (1, \frac{\pi}{2}]$.

Note. If $b < \delta(\gamma)$, then the distortion ropelength L/τ_b is always at least $2\delta(\gamma) \geq \pi$. To check this, take p and q with $\delta(p,q)$ close to $\delta(\gamma)$. Of course $d(p,q) \leq L/2$ and $|p-q| \geq \tau_b$, so dividing gives us $\delta(p,q) \leq L/2\tau_b$.

PROPOSITION 4.1. *The distortion thickness τ_b is upper semicontinuous in the C^0 topology on the set of closed rectifiable curves, for any fixed b.*

Proof. The claim is that if a sequence of curves γ^k approaches a limit γ^0, then their thicknesses $\tau^k := \tau_b(\gamma^k)$ satisfy

$$\overline{\lim}\ \tau^k \leq \tau^0.$$

If not, write $\overline{\lim}\ \tau^k = \tau^0 + 3\epsilon$, for some $\epsilon > 0$; passing to a subsequence, we may assume that $\tau^k > \tau^0 + 2\epsilon$ for all k.

Now, the infimum in the definition of τ^0 may not be realized, but we can certainly find a pair of points (p,q) which come close to realizing τ^0. That is, $|p-q|_0$, their straight-line distance on γ^0, is within ϵ of τ^0, while $\delta^0(p,q)$, their distortion on γ^0, is greater than b. So we have

$$\tau^k > |p-q|_0 + \epsilon, \qquad \delta^0(p,q) > b.$$

Since $|p-q|_k \to |p-q|_0$, but τ^k stays greater than this distance, we must have that (p,q) is not in contention for τ^k for all large enough k, that is, $\delta_k(p,q) \leq b$. On the other hand, the semicontinuity of δ means that $\liminf \delta_k(p,q) \geq \delta_0(p,q) > b$. This contradiction completes the proof. □

Note. The length of curves is lower semicontinuous (but not continuous) in this C^0 topology. Although we have discussed the existence of an arclength parameterization for each curve, we do not assume when discussing limits that constant-speed parameterizations are used. For instance, consider curves obtained by replacing one side of a square by finer and finer zigzags of twice the length. If each is parameterized by arclength, they approach a limit, which traces out the square at varying speed.

COROLLARY 4.2. *In the C^0 topology, the distortion ropelength of curves is lower semicontinuous.*

Proof. It is the quotient of a lower semicontinuous function (arclength) by an upper semicontinuous function (thickness). □

Let us consider some example curves. First take γ to be the unit circle. Given any $\phi \in [0, \frac{\pi}{2})$, any pair of points at angle 2ϕ is at distance $|p-q| = 2\sin\phi$, so it has distortion $\phi/\sin\phi$. Thus if $b = \phi/\sin\phi \in [1, \frac{\pi}{2})$ the thickness $\tau_b(\gamma)$ will be $2\sin\phi$. (The infimum here is not realized.)

Next, recall that the distortion of a wedge of (external) angle θ is $\sec\frac{\theta}{2}$. Given b, set $\theta = 2\sec^{-1} b$; if a curve γ has any corner sharper than this, its thickness $\tau_b(\gamma)$ will be zero. However, on a polygonal curve with no corners sharper than this, no pair of points on adjacent segments will be in contention for τ_b.

Finally, suppose γ bounds the convex hull of the unit circle and an exterior point (placed so that the angle at this point is θ). This curve has thickness $\tau_b = 0$ if $b < \sec\frac{\theta}{2}$, but if $b \geq \sec\frac{\theta}{2}$ its thickness is the same as that of the circle. This shows that the thickness is not C^0-continuous.

PROPOSITION 4.3. *Suppose γ is a $C^{1,1}$ curve. If the infimum defining τ_b is attained, this is at a doubly self-critical pair, so $\tau_b \geq c_2$. If not, but $\tau_b < \infty$, there is a pair of points with $\delta(p,q) = b$ and $|p-q| = \tau_b$.*

Proof. If the infimum is attained at (p,q) with $\delta(p,q) > b$, then all nearby pairs on γ also have $\delta > b$ and are in contention for τ_b, so since $|p-q|$ is a minimum, (p,q) must be a doubly self-critical pair. Otherwise, take a minimizing sequence (p_n, q_n), and pass to a convergent subsequence with some limit (p,q). Because the sequence was minimizing, $|p-q| = \tau_b(\gamma)$, with $\delta(p,q) \geq b$. But if $\delta(p,q) > b$, this pair would achieve the minimum, so instead $\delta(p,q) = b$. (Note also that this implies $d(p,q) = b\tau_b$.) □

Note. The minimum doubly self-critical distance c_2 can be achieved at a pair of arbitrarily small distortion $\delta > 1$, as if we put small hooks at the ends of a straight segment before continuing with a huge loop. Thus it is not clear in general that $\tau_b \leq c_2$. Perhaps, however, the arc between the points in a self-critical pair always has large distortion somewhere. Hence we might consider a refined definition of τ_b which allows in the infimum any (p,q) for which a subarc has distortion $\delta > b$.

5. The curvature thickness. Recall that $\sigma_k(\gamma) := \min\big(kr(\gamma), c_2(\gamma)\big)$, which is the smaller of the minimum doubly self-critical distance c_2 and the minimum radius of curvature r scaled by the factor k. This definition makes sense for any $k \geq 0$, but we are most interested in $k \in [1,2]$.

The special case of this definition with $k = 2$ was introduced in [LSDR]. This thickness σ_2 is characterized there as the maximum diameter of a tube around a C^2 curve γ for which the normal exponential map is an embedding. We will discuss this and similar notions in Section 6.

The further main results of [LSDR] also deal with σ_2 for C^2 curves and, translated into our notation, say that

1. If γ has ropelength $L/\sigma_2 < \frac{n\pi}{2}$ for some integer n, then there is some n-gon isotopic to γ (in the same knot class).
2. Thus, any nontrivial knot has ropelength at least $\frac{5\pi}{2}$.
3. If γ is a C^2 curve then its ropelength is at least twice its distortion: $L(\gamma)/\sigma_2(\gamma) \geq 2\delta(\gamma)$.
4. If γ is a C^2 curve, then $c_1(\gamma) \geq \sigma_2(\gamma)$. In other words, $c_1(\gamma) = c_2(\gamma)$ when either is less than $2r(\gamma)$

Note. This last result shows that $\sigma_k = \min(kr, c_1)$ for any $k \leq 2$, so that we could have defined σ_k in terms of c_1 instead of c_2.

Note. O'Hara [Oha1] attributes the definition of $c_1(\gamma)$ to Kuiper, and shows that if γ has $L/c_1 < n$ then it is isotopic to an n-gon. This is a weaker conclusion than the similar result of [LSDR], but requires no curvature bound.

Note. One of the authors has pointed out [Sul] that the discretization for thickness suggested by Stasiak [Sta] (and presumably used in numerical work reported in [KBM+]) actually discretizes σ_1. Here we approximate a smooth curve γ by a polygon with short edges of nearly equal length. Now look at solid cylinders of equal diameter around each edge. Increase the diameter until some nonadjacent cylinders first intersect; this gives an approximation of $\sigma_1(\gamma)$.

It is clear that $\sigma_k = 0$ for any curve with a sharp corner. Thus it is not possible to have any useful bound of the form $\sigma_k \geq C\tau_b$. However, we can bound τ_b below in terms of σ_k, because a curvature bound implies that short segments of a curve have small distortion. We use a lemma, which improves one from [Oha, p. 242] by a factor of two, and is in fact just a special case of Schur's Theorem (see [Che, p. 36]).

LEMMA 5.1. *If $\kappa(\gamma) \leq 1$, then for any points p and q on γ with $d(p,q) = d \leq 2\pi$, we have $|p-q| \geq 2\sin d/2$, or equivalently,*

$$\delta(p,q) \leq \frac{d/2}{\sin d/2}.$$

Proof. Let x be the point halfway along the arc from p to q. Orient the curve so that the tangent vector is vertical at x. At arclength s from x, the curve has turned less than angle s, so the vertical component of the unit tangent vector is still at least $\cos s$. Thus out to arclength $d/2$, the vertical component of the difference vector $x-p$ or $q-x$ is at least $\int_0^{d/2} \cos s\, ds = \sin d/2$. Thus $|p-q| \geq 2\sin d/2$, as desired. □

DEFINITION 5.1. *For any $b \in [1, \frac{\pi}{2}]$, define k_b to be the unique solution of the equivalent equations*

$$b = \frac{\arcsin k_b/2}{k_b/2}, \qquad \frac{bk_b/2}{\sin bk_b/2} = b.$$

Note that k_b increases monotonically from 0 to 2 as b increases from 1 to $\frac{\pi}{2}$, and thus $bk_b \leq \pi$.

THEOREM 5.2. *For any $1 \leq b \leq \frac{\pi}{2}$ and any closed curve γ, we have*

$$\tau_b(\gamma) \geq \sigma_{k_b}(\gamma).$$

Proof. Write $k = k_b$. If $\sigma_k = 0$ (as when γ is not $C^{1,1}$) there is nothing to prove. Otherwise, rescale so that $\sigma_k(\gamma) = k$. Then we have $c_2(\gamma) \geq k$ and $\kappa \leq 1$ everywhere on γ. Now apply Proposition 4.3; if the infimum defining τ_b is attained, $\tau_b \geq c_2 \geq k$, and we are done.

Otherwise, the proposition gives us points p and q with $|p-q| = \tau_b$ and $d(p,q) = b\tau_b$. If $\tau_b < k$ we want to use the lemma to derive a contradiction. Certainly then $d(p,q) < kb \leq \pi$ so the lemma applies, giving

$$b = \delta(p,q) \leq \frac{b\tau_b/2}{\sin b\tau_b/2}.$$

But $\frac{x}{\sin x}$ is an increasing function, so since $b\tau_b < bk$, we get $b < \frac{bk/2}{\sin bk/2}$ contradicting the definition of k_b. □

In particular, we have $\tau_{\frac{\pi}{3}} \geq \sigma_1$ and $\tau_{\frac{\pi}{2}} \geq \sigma_2$. Since $\tau_b(\gamma)$ and $\sigma_k(\gamma)$ are nondecreasing in b and k respectively, it follows that $\tau_b \geq \sigma_k$ whenever $b > 2\arcsin(k/2)/k$. We can also get bounds for other b and k.

COROLLARY 5.3. *For any $b \in [1, \frac{\pi}{2}]$, $k \geq k_b$ and any curve γ,*

$$\tau_b(\gamma) \geq \frac{k_b}{k}\sigma_k(\gamma).$$

In particular, $\tau_{\frac{\pi}{3}} \geq \sigma_k/k$ for $k \geq 1$.

Proof. Apply the theorem to b and k_b, and use the fact that by definition of σ_k, we have $\sigma_{k_b} \leq \sigma_k \leq \frac{k}{k_b}\sigma_{k_b}$. □

Note. Combining the result $\tau_{\frac{\pi}{2}} \geq \sigma_2$ with our previous note about distortion ropelength, we recover the result from [LSDR] that $L/\sigma_2 \geq 2\delta(\gamma)$. In fact, their proof is similar in spirit to our proof of Theorem 5.2.

Note. If γ is the unit circle, then for any b we have $\tau_b(\gamma) = \sigma_{k_b}(\gamma) = k_b$, so the theorem is sharp.

6. Thickness notions related to normal tubes. Other authors have suggested various notions of thickness defined as the largest diameter of a normal tube before some property fails. From [LSDR], we have seen that σ_2 is the infimum of diameters for which the normal exponential map fails to be an embedding.

Diao *et al.* [DEJvR] suggest that we require the $\frac{d}{2}$-neighborhood to be a solid torus and then look at the intersection of this neighborhood with each normal plane (assuming γ is C^1). If further we ask that the zero-component of each such intersection be a meridian disk in the solid torus, intersecting γ only at the original point, then the first d for which this fails is their C^1-continuous measure of thickness, which we call $t_D(\gamma)$.

O'Hara [Oha2, p. 60] suggests defining thickness simply as the infimum of all d such that the $\frac{d}{2}$-neighborhood of γ is not topologically a solid torus; we will call this $t_O(\gamma)$. However, the notion he actually uses is different. Let t_B be the infimal diameter of balls in space whose intersection with γ is not a single arc (and thus is disconnected or all of γ). O'Hara's implicit claim is that $t_B \leq t_O$.[1] Certainly we can have $t_B < t_O$, as for a polygon, where $t_B = 0$.

It is clear from these definitions that $\sigma_2 \leq t_D \leq t_O$, and that $t_D \leq c_1$. Combining these with the final result of [LSDR] shows that $t_D = c_1 = c_2$ whenever these are less than $2r(\gamma)$.

For the ellipse $(x/a)^2 + (y/b)^2 = 1$ with $a < b$, we have $c_2 = 2a = t_O$, $r = a^2/b < a$, and $c_1 = t_D$ with $\sigma_2 = 2r < t_D < t_O$. (Explicit values of τ_b here would involve elliptic integrals.) Suppose we now take the upper half of this ellipse, and attach it in a $C^{1,1}$ fashion to a large loop in the lower half plane; this new curve has $c_2 = 2a \ll t_O$.

Now consider a helix of radius ρ and pitch p, the intersection of a cylinder of radius ρ with the helicoid $z = pr\cos\theta$. (To get a closed curve, we could splice a long piece of this helix into a huge loop.) For steep pitch, this has no doubly self-critical points, but if $p^2 \leq -\min(\frac{\sin x}{x})$, we have $c_2 < 3\rho$, approaching 0 as the pitch decreases. If the pitch is shallow enough, $c_2 \leq 2\rho$ and then $t_O = c_2$, but for an intermediate range of pitches (when $c_2 > 2\rho$ but is still finite), we again find $t_O >> c_2$, because by the time the normal tube is thick enough to see the self-critical points, it already extends across the axis of the cylinder, so it is still homotopy equivalent to the helix.

7. Shortest curves in a knot class. We have made some progress in understanding the structure of curves minimizing the distortion ropelength, and believe the following:

CONJECTURE 7.1. *For each nontrivial knot type K there exists a shortest curve $\gamma \subset \mathbb{R}^3$ of fixed distortion thickness in that knot type; that is, each knot type has a curve γ of shortest distortion ropelength. Any such γ has bounded curvature, and thus is of smoothness class $C^{1,1}$. Away from doubly self-critical points realizing the thickness, γ must be straight. If we scale to make the distortion thickness $\tau_{\frac{\pi}{2}}(\gamma) = 1$, then the curvature of γ is bounded by 1, so that also the curvature thickness $\sigma_k(\gamma) = 1$ for all $k \geq 1$. Finally, for any $b \geq \pi/3$, we have $\tau_b(\gamma) = 1$, so all our notions of thickness agree for a ropelength minimizer γ.*

Note. It seems that the ropelength minimizer for the Hopf link must be the obvious candidate: two unit circles in perpendicular planes passing through each others' centers. For this link, $r = c_1 = c_2 = 1$, so $\sigma_k = 1$ for any $k \geq 1$. Also, the distortion thickness for any $b \geq \frac{\pi}{3}$ is $\tau_b = 1$. If we extend the definitions of t_D, t_O and t_B to links in the obvious ways, these also all equal 1. Presumably, in the ropelength minimizer for the connected sum of two Hopf links, the middle component (linking both others) will be

[1] In fact, for any diameter greater than t_O, he claimed to get a ball with disconnected intersection with γ. This can never be true if γ is a circle, so we have tried to correct for this in our definition of t_B.

a stadium curve, built of two semicircles and two straight segments; this shows we cannot expect minimizers to be C^2.

REFERENCES

[Che] S.S. CHERN, Curves and surfaces in Euclidean space, In S.S. Chern, editor, *Studies in Global Geometry and Analysis*, 16–56, Math. Assoc. Amer., 1967.

[DEJvR] YUANAN DIAO, CLAUS ERNST, AND E.J. JANSE VAN RENSBURG, Energies of knots, Preprint, 1996.

[FHW] MICHAEL H. FREEDMAN, ZHENG-XU HE, AND ZHENGHAN WANG, On the Möbius energy of knots and unknots, *Annals of Math.* **139** (1994), 1–50.

[Gro1] MIKHAEL GROMOV, Homotopical effects of dilatation, *J. Differential Geometry* **13** (1978), 303–310.

[Gro2] MIKHAEL GROMOV, Filling Riemannian manifolds, *J. Differential Geometry* **18** (1983), 1–147.

[GLP] MIKHAEL GROMOV, J. LAFONTAINE, AND P. PANSU, *Structures métriques pour les variétés riemanniennes*, Cedic/Fernand Nathan, Paris, 1981.

[JvR] E.J. JANSE VAN RENSBURG, Personal communication, June 1996.

[KBM+] VSEVOLOD KATRITCH, JAN BEDNAR, DIDIER MICHOUD, ROBERT G. SCHAREIN, JACQUES DUBOCHET, AND ANDRZEJ STASIAK, Geometry and physics of knots, *Nature* **384** (November 1996), 142–145.

[KS] ROBERT B. KUSNER AND JOHN M. SULLIVAN, Möbius energies for knots and links, surfaces and submanifolds, In Willam H. Kazez, editor, *Geometric Topology*, 570–604, Amer. Math. Soc./International Press, 1997. Proceedings of the Georgia Int'l Topology Conference, August 1993.

[LSDR] RICHARD A. LITHERLAND, JON SIMON, OGUZ DURUMERIC, AND ERIC RAWDON, Thickness of knots, Preprint, 1996.

[Oha] JUN O'HARA, Energy of a knot, *Topology* **30** (1991), 241–247.

[Oha1] JUN O'HARA, Family of energy functionals of knots, *Topology Appl.* **48** (1992), 147–161.

[Oha2] JUN O'HARA, Energy functionals of knots II, *Topology Appl.* **56** (1994), 45–61.

[Sta] ANDRZEJ STASIAK, Lecture at the AMS special session on *Physical Knot Theory*, Iowa City, March 1996.

[Sul] JOHN M. SULLIVAN, Lecture at the IMA workshop on *Topology and Geometry in Polymer Science*, June 1996.

[Tor] TATIANA TORO, Lecture at the IMA workshop on *Topics Related to Nonlinear PDE*, March 1996.

PERCOLATION OF LINKED CIRCLES

Y. DIAO* AND E.J. JANSE VAN RENSBURG†

Abstract. Let P_δ be a Poission process with density δ in $\mathcal{R}^3$, and let each point in the Poisson process be the center of a unit circle with normal vector a uniform random variable on the 2-sphere. Is there a critical density of the Poisson process such that there exists an infinite unsplittable link of circles? Clearly, if $\delta = 0$, then no such infinite unsplittable link exists, and if δ approaches ∞, then it becomes likely that an infinite unsplittable link exists. We prove that there exists an infinite unsplittable link at a finite value of δ. Similar results are true for other models of random closed curves that do not have the volume exclusion property.

Key words. Knots, Links, Knotted Polygons, Percolation.

1. Introduction. Topological entanglement (knotting and linking) and its physical ramifications arise naturally in science. For example, physical properties of synthetic high polymers depend on entanglements of polymer strands [1] and molecules which representing chiral knots are optically active [1,2]. Self-entanglements in linear polymers may be trapped as "tight knots" by crystallization [3,4] and this phenomenon is expected to have long-term effects on the behaviour of polymer melts. A random cyclization reaction on a dilute solution of polymers of equal molecular weight [5,6] generates a random family of polymer circles, which may be linked.

In this paper, we will be considering the linking of randomly placed closed curves in $\mathcal{R}^3$. In particular, what is the probability that there is an infinite unsplittable link of curves? If the density of curves is low, then it may be expected that there exists no infinite unsplittable link, but as the density is increased, an infinite unsplittable link is inevitable. At the critical value of the density, beyond which there is a positive probability that an infinite unsplittable link exists, we say that the curves percolate. We study this question in the case of geometric circles with centres generated by a Poisson process. Our methods will also apply to other cases so long as the random closed curves involved do not have the volume exclusion property. We choose the circles mainly for the simplicity in stating the theorems and proofs.

The study of random knotting and linking originated in the Frish-Wasserman-Delbrück conjecture: The probability that a randomly embedded circle is knotted tends to one as the length of the circle tends to infinity [7, 5]. This conjecture has been proven for self-avoiding polygons in the lat-

* Department of Mathematics, University of North Carolina at Charlotte, Charlotte, NC 28223, USA.

† Department of Mathematics and Statistics, York University, North York, Ontario, M3J 1P3, Canada.

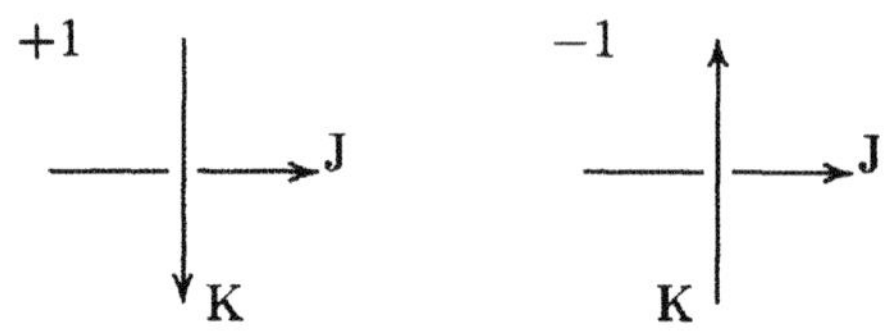

FIG. 1. *Sign convention.*

tice by Sumners and Whittington [8], and independently by Pippenger [9]. A similar result is true for Gaussian random polygons and equilateral random polygons in $\mathcal{R}^3$ [10,11]. The linking of two random curves curves were studied in [12,13], and some applicable formulae were obtained.

2. Linking of two randomly placed circles. Two disjoint simple closed curves J and K form an unsplittable link if no topological 2-sphere separates them. Thus, if any topological 2-sphere (S^2) is disjoint with $J \cup K$, then $J \cup K$ is contained in one of the two path components of $\mathbf{R}^3 \setminus S^2$. Assign arbitrary orientations to J and K and consider a plane projection of the link $J \cup K$. The projection is regular if all multiple points are double in the projection and the double points are the result of projected arcs which cross transversely. The projection becomes a link diagram if a small arc is removed from the underpassing arc about the double point in the projection. Examples of projected double points in a link diagram is shown in Figure 1. The crossovers in Figure 1 can be signed by using the orientation of the circles and by following the convention in Figure 1. The (Gaussian) linking number (or homological linking number) of the curves J and K (with respect to the assigned orientation) is defined to be one-half the sum of all the signs on crossings *between* J and K. It is denoted by $\ell(J, K)$. The linking number is a topological invariant of $J \cup K$, and if $\ell(J, K) \neq 0$, then $J \cup K$ is an unsplittable link [14,15].

We consider the linking of unit, planar (geometric) circles in a box of sides R. Consider first the linking of just 2 circles with centers a distance $2r$ apart, and randomly oriented (the normal unit vectors of the circles are independent uniform random variables on the 2-sphere). We call such circles "uniform random circles". The question here is to compute the probability that two uniform random circles, with centers a distance $2r$ apart, are linked. We sum the result up in theorem 1:

THEOREM 2.1. *The probability that two independent uniform random circles are linked, given that their centers are a distance $2r$ apart, is*

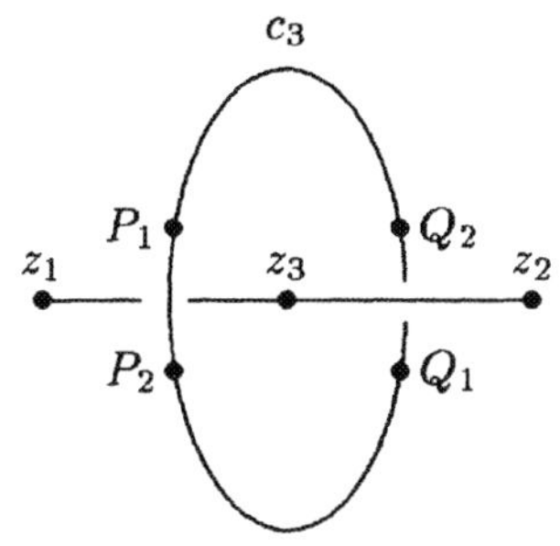

FIG. 2. *P_1 and Q_1 separate P_2 and Q_2 on c_3.*

(2.1)
$$P(r) = \begin{cases} 1 - r, & \text{for } 0 \leq r \leq 1, \\ 0, & \text{for } r > 1. \end{cases}$$

Remark: If $f(r)$ is the probability density that the centers of two uniform random circles are a distance $2r$ apart, then the probability that the two circles are linked is

(2.2)
$$P = \int_0^1 (1 - r) f(r) dr.$$

Proof: Let c_1 and c_2 be two uniform random circles and let z_1 and z_2 be the centers of c_1 and c_2, respectively. $|z_1 - z_2| = 2r$ is fixed, where $|z_1 - z_2|$ is the Euclidean distance between z_1 and z_2. Evidently, if $r \geq 1$, then c_1 and c_2 are always unlinked. So assume that $0 < r < 1$. The two unit spheres centered at z_1 and z_2, respectively, then intersect at a circle of radius $r_1 = \sqrt{1 - r^2}$. Call this circle c_3 and let z_3 be its center. It is obvious that c_1 and c_2 are unsplittable if and only if they both intersect c_3 in such a way that the intersection points P_1, Q_1 of c_1 with c_3 and the intersection points P_2, Q_2 of c_2 with c_3 separate each other as shown in Figure 2.

Now, let $\mathbf{u}$, $\mathbf{v}$ be the normal vectors of c_1 and c_2, respectively, and let θ, ϕ be the smaller angles between $\mathbf{u}$, $\mathbf{v}$ and the straight line through z_1 and z_2, respectively. See Figure 3. If c_1 is to intersect c_3 at two points, then $\sin^{-1} r < \theta \leq \frac{\pi}{2}$. θ is also a random variable and its probability density function is $\sin\theta$. A similar discussion also holds for ϕ.

Now fix P_1 and Q_1 on c_3 first (so θ is also fixed). If $\phi \geq \theta$, then in order that c_1 and c_2 are unsplittable, P_2 can only move on two separated arcs of equal length $2r_1 \cos^{-1}(\frac{r}{r_1} \cot\theta)$, ($P_2$, Q_2 are oriented according to

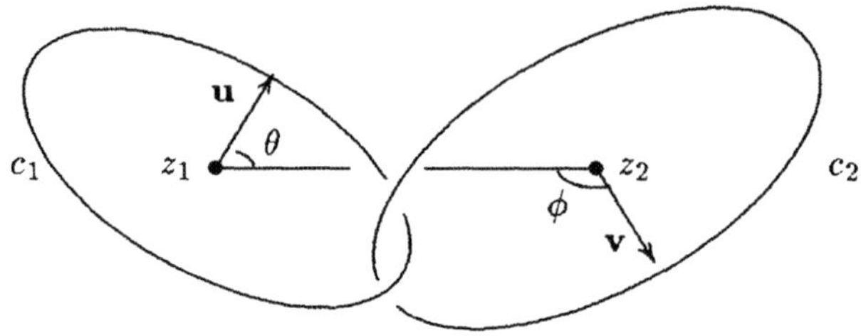

FIG. 3. *The positions in which c_1 and c_2 are unsplittable.*

the orientation of c_3, see Figure 3). Similarly, if $\sin^{-1} r < \phi \leq \theta$ then P_2 can only move along two separated arcs of equal length $2r_1 \cos^{-1}(\frac{r}{r_1} \cot \phi)$. This leads to the following formula:

$$
\begin{aligned}
&P(c_1 \cup c_2 \text{ is unsplittable}) \\
&= \int_{\sin^{-1} r}^{\frac{\pi}{2}} \sin\theta d\theta \{ \int_{\theta}^{\frac{\pi}{2}} \sin\phi \cdot \frac{2}{\pi} \cos^{-1}(\frac{r}{r_1} \cot\theta) d\phi \\
&+ \int_{\sin^{-1} r}^{\theta} \sin\phi \cdot \frac{2}{\pi} \cos^{-1}(\frac{r}{r_1} \cot\phi) d\phi \}.
\end{aligned} \tag{2.3}
$$

By exchanging the integration order in the second integral of the above and exchanging the notation of θ and ϕ, (2.4) simplifies to

$$
\begin{aligned}
&\frac{4}{\pi} \int_{\sin^{-1} r}^{\frac{\pi}{2}} \sin\theta \cos^{-1}(\frac{r}{r_1} \cot\theta) d\theta \int_{\theta}^{\frac{\pi}{2}} \sin\phi d\phi \\
&= \frac{4}{\pi} \int_{\sin^{-1} r}^{\frac{\pi}{2}} \sin\theta \cos\theta \cos^{-1}(\frac{r}{\sqrt{1-r^2}} \cot\theta) d\theta.
\end{aligned} \tag{2.4}
$$

After letting $y = \sin^2\theta$ and integrating by parts, the above is simplified to $1 - r$. □

Let $B(R)$ be a box of side-length R ($[0, R] \times [0, R] \times [0, R]$). Suppose that the centers of two uniform random circles are independent uniform random variables in $B(R)$. We call the circles now "uniform random circles rooted in $B(R)$". Let r be the half distance between two uniform random circles rooted in $B(R)$, and let $f(r)$ be the probability density of r in $B(R)$. $f(r)$ is bound from below by the probability density of the outscribed sphere of $B(R)$ (with radius $R/2$, and thus $f(r) \geq \frac{6r^4}{R^6}(2^4 R^3 - 3 \cdot 2^4 R^2 r + 2^6 r^3)$). The probability that two random circles are linked can be calculated from $f(r)$ using theorem 1 and equation (2.2), and is seen to be positive, whatever the placements of the circles in $B(R)$.

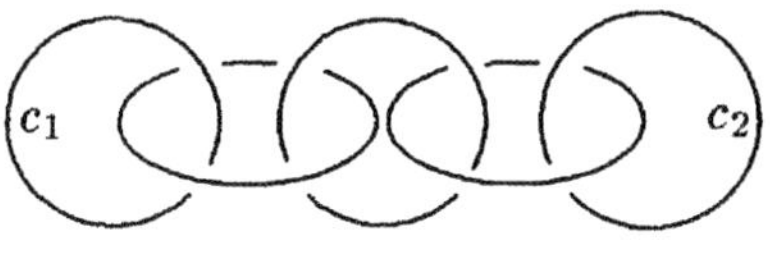

FIG. 4. *Two circles that are c-linked.*

Let c_1 and c_2 be any two (fixed) uniform random circles rooted in $B(R)$. c_1 and c_2 can be made part of an unsplittable link by adding a chain of $\lfloor \frac{\sqrt{3}}{2} R \rfloor + 2$ circles for one to the other. This is shown in Figure 4. We say that c_1 and c_2 are *c-linked.* The probability that c_1 and c_2 are c-linked by $\lfloor \frac{\sqrt{3}}{2} R \rfloor + 2$ uniform random circles is greater than a fixed positive constant, independent of their placements. This follows from the fact that the probability that c_1 and c_2 be c-linked is a continuous function of the positions of c_1 and c_2, which is a compact subset of $\mathcal{R}^{12}$ (in fact, it is $B(R) \times S^2 \times B(R) \times S^2$).

THEOREM 2.2. *Let c_1, c_2, ..., c_n be n uniform random circles rooted in $B(R)$, then they are disjoint with probability 1 and there exists a constant $\alpha > 0$ such that*

$$P(\cup_{i=1}^{n} c_i \text{ is unsplittable}) \geq 1 - e^{-\alpha n}. \tag{2.5}$$

Proof: Given $\{c_i\}_{i=1}^{n}$, it suffices to show that each c_i is c-linked to, say, c_1, with probability at least $1 - e^{-\alpha n}$ for some constant $\alpha > 0$. Consider the probability that c_1 and c_2 are c-linked. Divide c_3, c_4, ..., c_n into groups so that each group has $\lfloor \frac{\sqrt{3}}{2} R \rfloor + 2$ elements. If $\lfloor \frac{\sqrt{3}}{2} R \rfloor + 2$ does not divide n, disregard the last group which has less than $\lfloor \frac{\sqrt{3}}{2} R \rfloor + 2$ elements. Let m be the number of such groups. We have $m = \lfloor \frac{n-2}{\lfloor \frac{\sqrt{3}}{2} R \rfloor + 2} \rfloor > an$ for some positive constant $a < 1$. The probability that c_1 and c_2 are c-linked by any one of these groups is greater than a positive constant, say b, no matter what their positions are. Assuming that these two circles are fixed, then the event that they are c-linked by a particular group is an independent event of the being c-linked by another group. Thus, the probability (given c_1 and c_2 fixed) that c_1 and c_2 are not c-linked is less that $(1-b)^m$. But this probability is independent of the positions of c_1 and c_2, thus the probability that c_1 and c_2 are not c-linked is also at most $(1 - b)^m$. The same result applies to c_1 and c_i, $i = 3, 4, \ldots n$. Thus

$$P(c_1,\ c_i \text{ are not c-linked for some } i) \leq (n-1)(1-b)^m < e^{-\alpha n} \tag{2.6}$$

for some constant $\alpha > 0$. $\square$

3. Percolation of linked curves. In this section we study the percolation of linked circles. We assume that the centers of uniform random circles are generated by a Poission process in $\mathcal{R}^3$ of density (or intensity) δ. Several questions can be asked: Is there a critical value of the density (of the Poission process) where an infinite unsplittable link appears with a positive probability for densities greater than this critical value? Is this link finite in extent (with infinitely many circles), or is it infinite in extent? Suppose that an infinite cluster appears at a density (greater than the critical density), is there then only one such cluster? (In other words, is the infinite cluster unique?). We approach this problem by mapping it to the cubic lattice, where we will use results from bond percolation to construct a proof that a critical density exists.

Suppose that uniform random circles have centers generated by a Poission process of density δ in $\mathcal{R}^3$, and let the probability measure of this process be P_δ. Let T be a unit cube, we will define $W(T)$ to be the largest unsplittable link which contains a circle centered at a point in T, and we will denote the number of circles in this cluster by $|W(T)|$. Let $P_\delta(|W(T)| = \infty)$ be the probability that this unsplittable link is infinite at density δ. We define this probability as the *percolation probability*, and we denote it by

$$\rho(\delta) = P_\delta(|W(T)| = \infty). \tag{3.1}$$

The *critical percolation density* δ_c is then defined as

$$\delta_c = \sup\{\delta | \rho(\delta) = 0\}. \tag{3.2}$$

It is important to note that $\rho(\delta)$ is independent of T by the symmetric property of Poisson process. We can thus consider any T at our convenience.

THEOREM 3.1. *There exists a δ_c, $0 < \delta_c < \infty$, such that*

$$\rho(\delta) \begin{cases} = 0, & \text{if } \delta < \delta_c, \\ > 0, & \text{if } \delta > \delta_c. \end{cases} \tag{3.3}$$

Proof: It is obvious that $\rho(\delta)$ is a non-decreasing function of δ since increasing the density increases the number of circles in $\mathcal{R}^3$ and that can only increase the chance of linking. Thus it suffices to show that there exists a large δ such that $\rho(\delta) > 0$ since it is obvious that $\rho(\delta) = 0$ when δ approaches zero.

Since the Poisson process generates the centers of uniform random circles in $\mathcal{R}^3$ with a density δ, the probability that there are exactly k circles with their centers in a measurable volume of size v is $\frac{(\delta v)^k}{k!}e^{-\delta v}$. Partition $\mathcal{R}^3$ into unit cubes centered at points with integer coordinates.

These center points are also called lattice points of $\mathcal{Z}^3$. The probability that a given cube T contains exactly k circles is

$$\frac{\delta^k}{k!}e^{-\delta}. \tag{3.4}$$

Add the edges between adjacent lattice points (this corresponds to adding the edges joining the midpoints of adjacent cubes in $\mathcal{R}^3$). Let T_0 and T_1 be two adjacent cubes. The entire Poission process can be mapped to the lattice by calling the edge between the midpoints of T_0 and T_1 (denoted by e_1) "on" if the circles in $T_0 \cup T_1$ form an unsplittable link; otherwise we say e_1 is "off". The probability that an edge is "on" can be computed from equation (3.4) and from theorem 2. (We have to extend theorem 2 to rectangular boxes, but the argument goes through *mutatis mutandis*, and the result is the same). Since the probability that there are k circles in $T_0 \cup T_1$ can be computed from (3.4), and since there exists an $\alpha_1 > 0$ such that the k circles is an unsplittable link is bounded from below by $1 - e^{-\alpha_1 k}$, we obtain

$$P(e_1 \text{ is on}) \geq \sum_{k=0}^{\infty}(1 - e^{-\alpha_1 k})\frac{(2\delta)^k}{k!}e^{-2\delta} = 1 - e^{-2\delta(1-e^{-\alpha_1})}. \tag{3.5}$$

Thus, the probability that any particular edge is "on" approaches 1 as the density of the Poission process increases to infinity. The Poission process becomes a bond percolation process in $\mathcal{Z}^3$, but the event that a given edge is "on" is not independent from the states of the neighbouring edges, so we do not have a standard bond percolation process. Thus, we cannot directly apply results from bond percolation. To avoid this problem, we simplify the process by considering a 2 dimensional section of it, and map that to bond percolation in the square lattice $\mathcal{Z}^2$. Thus, consider a realization of the Poisson process of circles, and choose an infinite array of cubes as before but with their centers in the xy-plane.

Let δ now be so large such that the probability that the circles in any two given adjacent unit cubes form an unsplittable link is at least $1 - \beta$. This is possible by equation (3.5) and the arguments leading to it. In fact, we can choose β as close to 0 as we want by letting δ go to ∞. Note that the probability that a given edge is on is independent of the states of edges not incident with it. Let O be the origin in the lattice $\mathcal{Z}^2$ and let $W(O)$ be the cluster of on edges connected to it by a path of on edges. $W(O)$ is the linked cluster rooted at O. Let $\mathbf{Z}^2$ be the square lattice dual to $\mathcal{Z}^2$, and let an edge in $\mathbf{Z}^2$ be on if it crosses an on edge in $\mathcal{Z}^2$. This defines a bond percolation process in the dual lattice. If $W(O)$ is finite, then it is bound by a polygon $\gamma(n)$ in $\mathbf{Z}^2$ containing O such that every edge in $\gamma(n)$ is off, and $\gamma(n)$ contains n edges. There are at most $4n3^{n-1}$ such polygons with n edges (see page 17 in [16]). Given $\gamma(n)$, the probability that the edges in it are all off is at most $(\sqrt{\beta})^n$, since there are $n/2$ disjoint edges in $\gamma(n)$, and these are independent. Thus, the probability that $W(O)$ is finite is bounded from above as

$$P(W(O) \text{ is finite}) \leq \Sigma_{n=1}^{\infty} 4n3^{n-1}\sqrt{\beta}^n = \Sigma_{n=1}^{\infty} \frac{4n}{3}(3\sqrt{\beta})^n < 1,$$

if β is small enough (and thus δ is large enough). Thus, $W(O)$ is infinite in size with a positive probability if δ is large enough (but still finite). □

δ_c is the critical density where a cluster of infinite extent occurs, in the sense that the percolation cluster in $\mathcal{Z}^3$ is infinite in extent. Let $E(X)$ be the expected value of the quantity X, then we define

$$\delta_T = \sup\{\delta | E(|W|) < \infty\}. \tag{3.6}$$

where W is a cluster starting from a given circle in a given cube. Thus, if $\delta < \delta_T$, then there are no unsplittable links with infinite number of circles, and if $\delta > \delta_T$, then there is an unsplittable link with an infinite number of circles. Thus, $\delta_T \leq \delta_c$, since an unsplittable link with an infinite number of circles may appear before there is an unsplittable link of infinite extent. We show that this does not happen:

THEOREM 3.2. $\delta_T = \delta_c$.

Proof: At a given density δ, the expected number of circles in a unit cube T_0 is given by

$$\sum_{k=0}^{\infty} k\frac{\delta^k}{k!}e^{-\delta} = \delta, \tag{3.7}$$

and this is finite. Thus, the expected value of the number of circles in a finite union of cubes is also finite. Consequently, an unsplittable link with an infinite number of circles can only appear if it is in an infinite union of cubes. Hence $\delta_T \geq \delta_c$. □

How many infinite unsplittable links are there at a given density of the Poission process? We have the following

THEOREM 3.3. *If $\delta > \delta_c$, then there exists either one or infinitely many infinite unsplittable links with probability one.*

Proof: Notice that the event that there exists an infinite unsplittable link does not depend on the states of any finite collection of random circles. Thus, if we let $\phi(\delta)$ be the probability of this event at density δ, then $\phi(\delta)$ satisfies the so called zero-one law (see [17]). That is, $\phi(\delta) = 0$ or $\phi(\delta) = 1$. If $\delta < \delta_c$, then

$$\phi(\delta) \leq \Sigma_T P_\delta(|W(T)| = \infty) = 0,$$

where in the above T runs through all cubes of side length 4 centered at points with coordinates divisible by 4. On the other hand, if $\delta > \delta_c$, then

$$\phi(\delta) \geq P_\delta(|W(T_0)| = \infty) > 0,$$

hence $\phi(\delta) = 1$. To see that there is either 1 or infinitely many unsplittable links in the case, we adapt an argument due to Zuev and Siderenko [18]. Suppose that $\delta > \delta_c$ and suppose that there are k infinite clusters, $k > 1$ and $k \neq \infty$. Consider any two of these, and select a circle on each. These circles are a finite distance apart, and we can link the two infinite clusters by adding a finite chain of linked circles between the selected circles. This construction does not change the density of the Poisson process, since we only add a finite number of circles. But then k has reduced by 1. This gives a contradiction, and so there are either 1 or infinitely many unsplittable links. □

It is unclear whether the above can be strengthen to show uniqueness of the infinite cluster.

4. Conclusions. In this paper we have considered the percolation of linked circles in $\mathcal{R}^3$. We have shown that there is a positive probability that a set of circles in a box of side-length R will form an unsplittable link. This result was used to study the percolation of links in $\mathcal{R}^3$. Assuming that the centers of the circles are generated by a Poission process of density δ, we proved that there exists a critical value of the density, $0 < \delta_c < \infty$, such that if $\delta < \delta_c$, then there is no unsplittable infinite link. If $\delta > \delta_c$, then there exists an infinite unsplittable link, infinite in extent.

These results can be extended to other models of random closed curves such as equilateral random polygons, as long as these random closed curves do not have the volume exclusion property. This is obvious if the random curves are all independent of each other and are identical in shape (so they can be treated just as the circles). If the shapes change (such as the polygons), then the proofs need to be modified. For instance, if the random curves are equilateral random polygons of odd number of edges, then any two of them (with their starting points in a fixed volume such as a cube) can be linked with a positive probability. This probability is greater than a constant that is independent of the shapes of the polygons. In this case, all the theorems and proofs on the circles can be applied with little adjustments. But if the polygons are of even length, then it can happen that a polygon can be very close to a shape such as a doubled line segment, making it very hard to link with other polygons. In this case, it is unlikely that all random curves in one cube can form an unsplittable link with high probability, even if the density of the Poisson process goes to infinity. We will then have to change the definition of an edge to be "on" to something slightly weaker. This is left to our reader to explore.

Acknowledgments. The authors are grateful to the Institute for Mathematics and its Applications for financial support.

REFERENCES

[1] H.L. Frisch and D. Klempner, *Topological Isomerism and Macromolecules,* Advances in Macromolecular Chemistry V2 , W.M. Pasika, ed., Academic Press (1970).

[2] R.L. Kornegay, H.L. Frisch and E. Wasserman, *A Calculation of the Optical Activity of a Trefoil Knot,* J. Org. Chem. 34 (1969), 2030.

[3] P.G. de Gennes, *Tight Knots,* Macromolecules 17 (1984), 703–704.

[4] R.C. Lacher, J.L. Bryant, L.N. Howard and D.W. Sumners, *Linking Phenomena in the Amorphous Phase of Semicrystalline Polymers,* Macromolecules 19 (1986), 2639–2643.

[5] H.L. Frisch and E. Wasserman, *Chemical Topology,* J. Am. Chem. Soc. 83 (1961), 3789–3795.

[6] J. Roovers and P.M. Toporowski, *Synthesis of High Molecular Weight Ring Polys-tyrenes,* Macromolecules 16 (1983), 843–849.

[7] M. Delbrück, *Knotting Problems in Biology,* Proc. Symp. Appl. Math. 14, American Mathematical Society (1962), 55–63.

[8] D.W. Sumners and S.G. Whittington, *Knots in Self-Avoiding Walks,* J. Phys. A: Math. Gen. 21 (1988), 1689–1694.

[9] N. Pippenger, *Knots in Random Walks,* Discrete Appl. Math. 25 (1989), 273–278.

[10] Y. Diao, *On Knotting of Randomly Embedded N-gons in* $\mathbf{R}^3$, Ph. D. Thesis, Florida State University, 1990.

[11] Y. Diao, N. Pippenger and D.W. Sumners, *On Random Knots,* J. of Knot Theory and its Ramification, in press.

[12] J. des Cloizeaux and R. Ball, *Rigid Curves at Random Positions and Linking Numbers,* Commun. Math. Phys. 80 (1981), 543–553.

[13] W.F. Pohl, *The Probability of Linking of Random Closed Curves,* 113–126.

[14] G. Burde and H. Zieschang, *Knots,* de Gruyter, New York 1985.

[15] D. Rolfsen, *Knots and Links,* Publish or Perish, Inc. Berkeley, 1976.

[16] G. Grimmett, *Percolation,* Springer-Verlag, New York, 1989.

[17] G. Grimmett and D. Stirzaker, *Probability and Random Processes,* Clarendon Press, Oxford, 1982.

[18] S.A. Zuev and A.F. Siderenko, *Continuous Models of Percolation Theory, II,* Teor. Mat. Fiz. 62 (1985), 253–262.

[19] Y. Diao, *Unsplittability of Random Links,* Journal of Knot Theory and its Ramifications, Vol. 3, No. 3 (1994), 379–389.

MINIMAL LINKS IN THE CUBIC LATTICE

R. UBERTI*, E.J. JANSE VAN RENSBURG*, E. ORLANDINI†, M.C. TESI‡, AND S.G. WHITTINGTON§

Abstract. How many edges are necessary and sufficient to realize two closed simple curves in the cubic lattice forming a link of a given type? Let this number be the edge number of the link. We show that the minimal edge number of links is subadditive with respect to the connected sum of links. In addition, we prove the existence of an edge index, which is the average number of edges per component under iterated connected sums of links of a given type. The minimal edge numbers of some simple links are estimated using simulated annealing.

1. Introduction. The *cubic lattice* $\mathcal{Z}^3$ is the set of points (*vertices*) with integer Cartesian coordinates in $\mathcal{R}^3$. This lattice can be augmented by the addition of all unit line segments (*edges*) between vertices a unit distance apart. A *polygon* in $\mathcal{Z}^3$ is an unlabeled vertex-avoiding closed path of edges (also called a cycle). A polygon with the least number of edges has length 4, since $\mathcal{Z}^3$ has girth 4. Two polygons are equivalent if they can be made identical by a translation, and we abuse our notation by referring to the resulting equivalence classes as polygons. By inclusion, every polygon is a tame, simple, closed piecewise linear curve in $\mathcal{R}^3$. The knot-type of a polygon, and the linking between polygons, are thus well defined. Two tame circles are *joined* if they are endpoints of a sequence of tame circles in which successive pairs are topologically linked. A set of tame circles form a *connected link* if every pair of tame circles in the set are joined. We will now refer to a "connected link" simply as a "link", and each circle in a link will be called a component. A link with n components of type L will be indicated by $L(n)$. The standard unit vectors in $\mathcal{Z}^3$ will be denoted by $\imath$, $\jmath$, k. We will often perform a lexicographic ordering of a set of points; we first order the vertices with respect to $\imath$, then with respect with $\jmath$ and then k. Thus, we shall refer to $\imath$ as the *first direction*, $\jmath$ as the *second direction* and k as the *third direction*.

It is known that any (tame) graph with maximum vertex degree six can be embedded in the cubic lattice; that is, there exists a cubic lattice representation for any such graph. In particular, there exists lattice representatives for any knot or link [1]. It can be checked that 8 edges are necessary to construct a non-trivial link consisting of two polygons (this will be the Hopf-link). It is also known that 24 edges are necessary to

* Department of Mathematics and Statistics, York University, 4700 Keel street, North York, Ontario, M3J 1P3, Canada.

† Department of Theoretical Physics, Oxford University, 1 Keble Road, Oxford, OX1 3NP, United Kingdom.

‡ Mathematical Institute, University of Oxford, Oxford, OX1 3LB, United Kingdom.

§ Department of Chemistry, University of Toronto, Toronto, M5S 1A1, Canada.

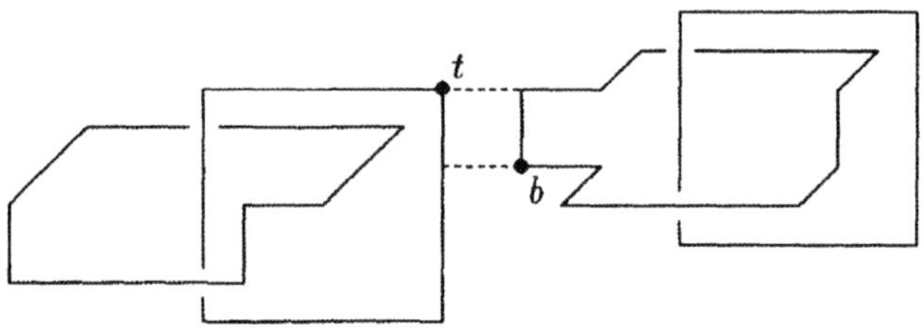

FIG. 1. *Concatenation of two lattice links. The link-type of the resulting lattice link is dependent on the lattice conformations of the initial lattice links.*

construct a polygon which is a non-trivial knot [2], and the only knot-type which can be constructed with 24 edges is the trefoil [3] (3_1 in Alexander-Briggs notation [4]). The minimum number of edges necessary to realize a knotted polygon of a given knot-type in the cubic lattice is the *edge number* of the knot-type [5]. We define the *edge number* of a link of type L to be the minimum number of edges necessary to realize a link of type L in the cubic lattice, analogous to the edge number for knots.

2. Edge numbers and edge indices. Let $M(L(n))$ be the edge number of a link of type $L(n)$ (in other words, there exists a realization of lattice polygons with a total of $M(L(n))$ edges, and with link type $L(n)$, *and*, there is no such realization if fewer than $M(L(n))$ edges are available).

2.1. Concatenation. Two realizations of lattice links of types $L_1(n)$ and $L_2(m)$ can be combined to give a link in the lattice of type $L_3(n+m-1)$ (which depends on the conformations of $L_1(n)$ and $L_2(m)$) by *concatenation*. This is illustrated in figure 1. The *bottom vertex* and *top vertex* of a lattice link are the first and last vertices in a lexicographic ordering of all the vertices in the link. It is always the case that at least one edge incident with the top vertex is perpendicular to $\imath$. From these edges, choose that edge with lexicographic least midpoint as the *top edge* [6]. Similarly, there is always at least one edge incident with the bottom vertex and perpendicular to $\imath$. From these edges, choose that edge with lexicographic least midpoint as the *bottom edge*. Let ω and ν be lattice realizations of a link of type $L_1(n)$, and of type $L_2(m)$. Let the top edge of ω be t_ω, and the bottom edge of ν be b_ν. Translate and rotate ω in $\mathcal{Z}^3$ such that $t_\omega + \imath = b_\nu$. Delete t_ω and b_ν and join their endpoints by two edges in order to generate a new lattice link of type $L_3(n+m-1)$, as illustrated in figure 1.

The operation of concatenation shows that in general, the edge number of a link is *subadditive*. We state this in lemma 2.1:

LEMMA 2.1. *If $L_3(n+m-1)$ is a link type which can be obtained by concatenating minimal lattice realizations of links of type $L_1(n)$ (with*

FIG. 2. *Lattice polygons with 4 or 6 edges are all a rotation of one of the above. Each polygon is the boundary of a disk (with folds indicated with dotted lines) which cannot be pierced by a sequence of edges in the cubic lattice, and are hence contractible to a point.*

$M(L_1(n))$ *edges) and* $L_2(m)$ *(with* $M(L_2(m))$ *edges), then*

$$M(L_3(n+m-1)) \leq M(L_1(n)) + M(L_2(m)).$$

Proof. If lattice realizations of $L_1(n)$ with $M(L_1(n))$ edges and $L_2(m)$ with $M(L_2(m))$ are concatenated, such that a lattice link of type $L_3(n+m-1)$ is obtained (by the hypothesis), then this realization has at least $M(L_3(n+m-1))$ edges. □

If ω is a polygon linked to any other polygon ν, then ω has length at least 8 edges. To see this, we enumerate all polygons with 4 or 6 edges (there are no polygons with odd length, since the cubic lattice is bipartite, and the girth of the cubic lattice is 4). These are illustrated in figure 4. To see that none of these can be linked to another polygon, simply note that each one is the boundary of a disk which cannot be pierced by another polygon.

These observations prove a second lemma:

LEMMA 2.2. *If* $L_2(n-1)$ *is a link which can be obtained from a link* $L_1(n)$ *by deleting* 1 *component of* $L_1(n)$*, then*

$$M(L_2(n-1)) + 8 \leq M(L_1(n)).$$

Proof. Let ω be a realization of a lattice link of type $L_1(n)$ with $M(L_1(n))$ edges. Any polygon in ω contains at least 8 edges, by the arguments above, and deleting one will give a link of type $L_2(n-1)$ with $M(L_1(n)) - 8$ edges. □

In particular, we note that $M(L(n)) \geq 8n$, for any link with $n \geq 2$ components.

2.2. Branched edge index. Let $L^B(n)$ be that link of n components, such that every pair of components is either a topological link of type L, or is not topologically linked. $M(L^B(n))$ is the minimum number of edges in a lattice realization of the link $L^B(n)$. We call it the n-th *branched minimal edge number* of L. The average number of edges per component in the n-th branched minimal edge number is $\frac{1}{n}M(L^B(n))$. We now show that this average approaches a limit as n increases:

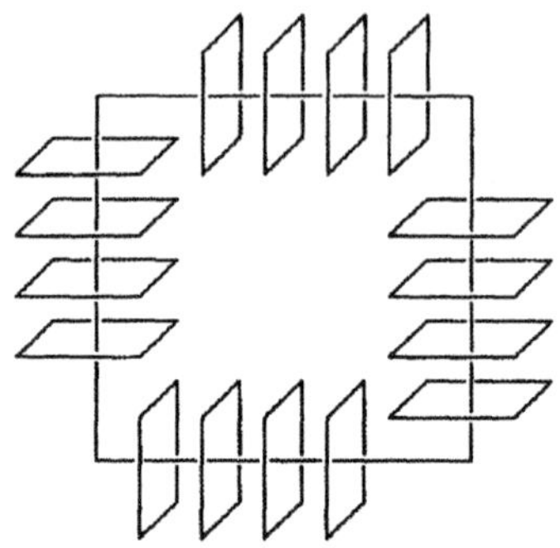

FIG. 3. *This construction can be used to compute bounds on the branched edge index of the Hopf-link. By increasing the side length of the large square, the number of edges per component goes to 9. Each of the smaller components is a square of length 8 edges.*

THEOREM 2.3. *The branched edge index of the link L is defined by the limit*

$$\mu_L^B = \lim_{n\to\infty} \frac{1}{n} M(L^B(n)) = \inf_{n\geq 2} \frac{1}{n-1} M(L^B(n)),$$

and moreover, $M(L^B(n)) \geq (n-1)\mu_L^B$, *where* $\mu_L^B \geq 8$ *for any link type L.*

Proof. If we concatenate two links $L^B(n)$ and $L^B(m)$, then by lemma 2.1:

$$M(L^B(n)) + M(L^B(m)) \geq M(L^B(n+m-1)),$$

since $L^B(n+m-1)$ is realized by the concatenation, but could be constructed with fewer than $M(L^B(n)) + M(L^B(m))$ edges. In other words, $M(L^B(n+1))$ is subadditive. But by lemma 2.2, $M(L^B(n)) \geq 8n$, and these facts are enough to prove existence of the limit and the bounds [7,8,9]. □

Figure 3 proves an upper bound on the branched edge index of the Hopf-link. Together with theorem 2.1 we obtain $8 \leq \mu_L^B \leq 9$ if L is the Hopf-link. Let $L_1(n)$ and $L_2(m)$ be two arbitrary links. We define the *connected sum*, $L_1(n)\#L_2(m)$, by taking the connected sum of a component of $L_1(n)$ with a component of $L_2(m)$. $L_1(n)\#L_2(m)$ is a link with $n+m-1$ components. The connected sum between some links is uniquely defined, in the sense that the outcome is independent of the components involved in the operation. For example, the connected sum between two Hopf-links (2_1^2 in Alexander-Briggs notation), or between the links 5_1^2 and 8_2^2 (figure 4), has a unique outcome. We can easily show this by noting that the components of both 5_1^2 and 8_2^2 can be interchanged by Reidemeister moves and a plane isotopy. In other words, if we colour one component red, and the other blue, then we can interchanged the colours by ambient isotopy in $\mathcal{R}^3$. We call links such as these *symmetric*.

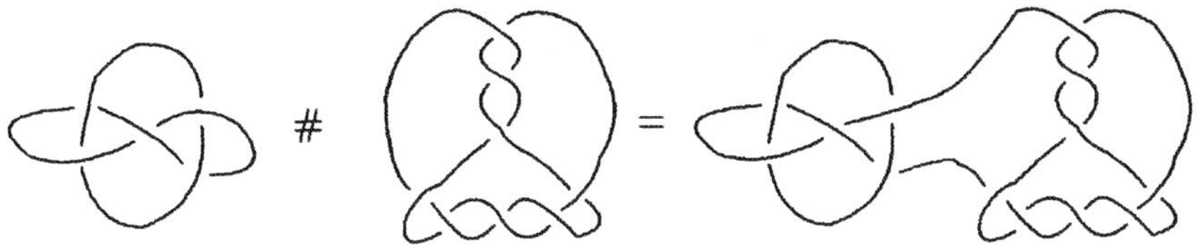

FIG. 4. *The connected sum of two symmetric links. The outcome is independent of the components involved.*

FIG. 5. *The link* 7_4^2.

In some cases, the outcome of a connected sum between two links is not unique. For example, any connected sum involving the link 7_4^2 (figure 5) depends on the component involved. This is also easily seen; note that one component in this link is an unknot, and the other a trefoil. Thus, if we colour the unknotted component red, and the knotted component blue, then we cannot interchange these by Reidemeister moves and a plane isotopy. A connected sum will necessarily involve either the blue or the red component. Links such as these will be called ***asymmetric***. (In fact, we define any link which is not symmetric to be asymmetric.)

Colour two components of a link with the same colour if and only if they can be interchanged by Reidemeister moves and a plane isotopy in a regular plane projection of the link. Then symmetric links will be 1-colourable (or monochromatic), while asymmetric links will require at least 2 colours. All the links in the standard link table up to and including 7_3^2 are monochromatic. The connected sum of two symmetric links is asymmetric, since one of the resulting 3 components will be linked to the 2 other components, but these are linked only to one other component. The link in figure 3 is 2-colourable, and thus asymmetric.

Let $L(2)$ be an asymmetric link, and colour its two components red and blue. Then there exist lattice realizations (not necessarily minimal) such that the bottom edge is either red or blue, and the top edge is either red or blue [1]. Consider 4 lattice realizations of $L(2)$, one with both top and bottom edges blue, say $\omega(bb)$, the second with both top and bottom edges red, $\omega(rr)$ and the third and fourth with differently coloured top and bottom edges, $\omega(br)$ and $\omega(rb)$. By adding edges in pairs to the top edge, we can also arrange matters such that all these have the same number of edges, say N. The colours of the edges can be preserved under concatenation if we can only concatenate two lattice links if the top and bottom edges are the same colour. By concatenating the lattice links ω above, we can construct a lattice link of type $L^C(n)$, and we call this the *bichromatic link of type* $L(2)$. In this link, every red component is linked only to blue components and every blue component is linked only to red components. Let the minimal edge number of this be $M(L^C(n))$, and we call this the *minimal bichromatic edge number* of the asymmetric link $L(2)$. Naturally, $M(L^C(n)) \geq M(L^B(n))$. The *bichromatic edge index* of an asymmetric link can be proven to exist as follows: Let τ_n and τ_m be minimal lattice representatives of the bichromatic links of type $L(2)$ with n and m components respectively. Suppose that the top edge of τ_n has colour x and the bottom edge of τ_m has colour y. Concatenate τ_n with $\omega(xy)$, then the new link has $n+1$ components and top edge of colour y, the same as the bottom edge of τ_m. Concatenate these now again to obtain a bichromatic link with $(n+m)$ components. (Here, x and y can be any of $\{b, r\}$). Thus

$$M(L^C(n)) + M(L^C(m)) + N \geq M(L^C(n+m)).$$

The same arguments as in theorem 2.1 thus imply theorem 2.2:

THEOREM 2.4. *For every asymmetric link $L(2)$ there exists a finite constant N such that the limit*

$$\mu_L^C = \lim_{n\to\infty} \frac{1}{n} M(L^C(n)) = \inf_{n\geq 1} \frac{1}{n}(M(L^C(n)) + N),$$

exists. μ_L^C is the bichromatic edge index of the asymmetric link $L(2)$. Moreover, $\mu_L^C \geq \mu_L^B$ and $M(L^C(n)) \geq n\mu_L^C - N$.

Theorem 2.2 can be generalized to m-colourable links, by adapting the arguments preceding it to asymmetric links which can only be coloured with m colours.

Let $L_{1,2}^B(n)$ be a link of n circles, where every pair of circles is the unlink, or is a link of type L_1 or a link of type L_2. Let $M(L_{1,2}^B(n))$ be defined as before. The *mixed branched edge index* of links of types L_1 and L_2 can now similarly be defined by noting that (under concatenation)

$$M(L_{1,2}^B(n)) + M(L_{1,2}^B(m)) \geq M(L_{1,2}^B(n+m-1)).$$

Thus

THEOREM 2.5. *The mixed branched edge index of the links of type L_1 and L_2 is defined by the limit*

$$\mu_{L_{1,2}}^B = \lim_{n\to\infty} \frac{1}{n} M(L_{1,2}^B(n)) = \inf_{n\geq 2} \frac{1}{n-1} M(L_{1,2}^B(n)).$$

TABLE 1
Minimal Edge Numbers for Links with 2 Components

$M(L)$	Links
16	2_1^2(8,8)
28	4_1^2(14,14)
34	5_1^2(22,12)
38	7_7^2(28,10)
40	6_1^2(22,18)
42	6_2^2(30,12) 6_3^2(30,12) 7_8^2(32,10)
44	7_6^2(32,12) 8_{15}^2(34,10)
46	7_1^2(26,20)
48	7_2^2(26,22) 7_3^2(36,12) 7_5^2(36,12) 8_{16}^2(36,12)
50	8_1^2(32,18)
52	7_4^2(42,10) 8_2^2(26,26) 8_5^2(36,16) 8_6^2(40,12) 8_{12}^2(40,12)
54	8_4^2(32,22) 8_7^2(38,16) 8_8^2(36,28) 8_{10}^2(42,12) 8_{13}^2(42,12) 8_{14}^2(40,14)
56	8_3^2(36,20) 8_{11}^2(44,12)
58	8_9^2(36,22)

Moreover, $\mu_{L_1}^B + \mu_{L_2}^B \geq 2\mu_{L_{1,2}}^B$.

Proof. The existence of $\mu_{L_{1,2}}^B$ follows from the subadditive relation above. To find the upper bound, concatenate minimal lattice links of types $L_1^B(n)$ and $L_2^B(n)$ to obtain $M(L_1^B(n)) + M(L_2^B(n)) \geq M(L_{1,2}^B(2n-1))$. Divide by n and let $n \to \infty$. □

Theorem 2.3 is easily generalized to include more than two link types.

3. Numerical results. The minimal edge numbers of links can be estimated by simulated annealing [10] implemented via the BFACF algorithm applied to linked polygons [11]. It can be shown that this algorithm is irreducible for linked polygons, by using exactly the same arguments as in reference [12]. We have computed an upper bound on minimal edge numbers for all links in the knot tables up to and including those with minimal crossing number 8. The results are in Table I. The first column indicates the minimal edge number, and the links are listed by name with the number of edges in each component following in brackets.

The number of different links of a given minimal edge number increases with the minimal edge number in Table I. If we count links with at most k components, then this increase is at most exponential. To see this, we argue as follows: Let $q_k(n)$ be number of distinct link-types, with at most k components, and with minimal edge number equal to n. Define $Q_k(n) = \sum_{m\leq n} q_k(m)$. Thus, $Q_k(n)$ is the number of distinct link-types with at most k components, and with minimal edge number *at most* n.

THEOREM 3.1. *$Q_k(n) \leq K^n$ for some $K > 0$. Moreover, there exists a real number, dependent on k, such that $\lim_{n\to\infty}[Q_k(n)]^{1/n} = \mathcal{Q}_k \geq 0$.*

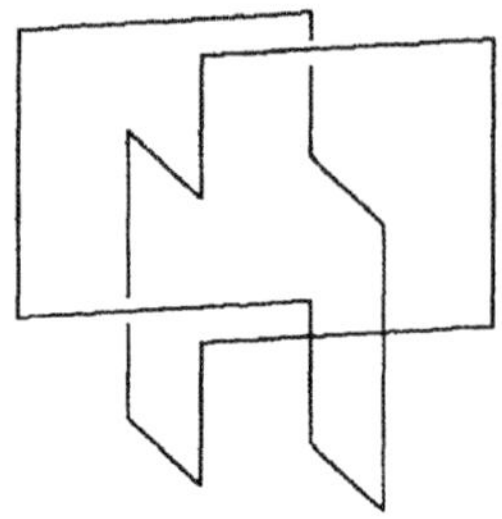

FIG. 6. *The link 4_1^2 represented as a minimal lattice link with 28 edges, 14 in each component. Observe that one component can be rotated into the other.*

Proof. Let ω be a lattice link with $l \leq k$ components. Then ω consists of k polygons with a total of n edges, which can be embedded in a cubic box of side length n to produce ω. The number of polygons with n edges, p_n, is known to be bound above by $2\mu^n$, where μ is the connective constant of the self-avoiding walk [13,14]. Since any polygon of n edges can be embedded in at most $(2n)^3$ ways to have a vertex in the box of side length n, an upper bound on $q_k(n)$ is $(2\mu^n)^k(2n)^{3k}$, which can be bound exponentially in n to produce an exponential bound on $Q_k(n)$. Since $Q_k(n)$ is non-decreasing, the existence of the limit follows. □

We were not able to prove a result equivalent to theorem 3.1 for links which may have arbitrary numbers of components. In particular, the number of distinct links with minimal edge number less than or equal to a n could grow as a factorial, but this seems not to be a likely situation. (The reason that the proof of theorem 3.1 fails when the number of components is not bounded is as follows: Consider a large square of side-length m in the square lattice, and count the number of ways that it can be linked to $\lfloor \epsilon m^2 \rfloor$ other squares of side-length m. This number diverges with m more or less as a factorial in m^2, even though the number of edges involved is only $O(m)$. The main problem is to show that all these realizations of lattice links can generate only exponentially many link-types - but this does not appear to be a simple problem.)

The results in Table I are upper bounds on μ_L^B, as expressed in theorem 2.3. In general, the minimal edge numbers tend to increase with minimal crossing number. As in the case with lattice knots, there are some exceptions to this rule of thumb. In Table I the link 7_7^2 occurs earlier than expected, as do 7_8^2 and 8_{15}^2. These are all non-alternating links. There are other interesting observations to be made from Table I. The $(2,k)$-torus links 2_1^2, 4_1^2, 6_1^2 and 8_1^2 appears to step up in increments of 12's and 10's. An argument in reference [5] applied to these would explain this observation partially (see also [15]). It is also apparent that symmetric links rarely distribute edges evenly between their components, at the minimal edge number. In Table I only the links 2_1^2, 4_1^2 and 8_2^2 have the same number

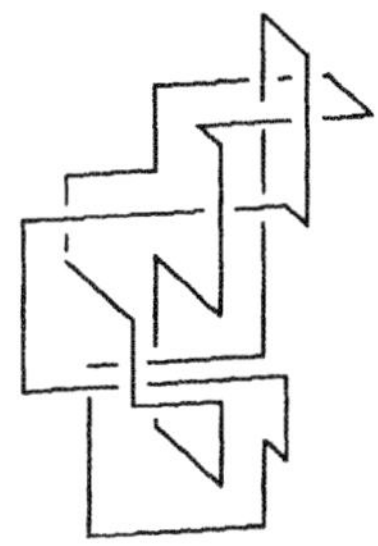

FIG. 7. *The link 8^2_2 represented as a minimal lattice link with 52 edges, 26 per component.*

TABLE 2
Minimal Edge Numbers for Links with 3 or 4 Components

$M(L)$	Links
34	6^3_3(14,10,10)
36	6^3_2(12,12,12) 7^3_1(22,12,12)
40	8^4_3(10,10,10,10)
42	6^3_1(16,14,12)
48	8^4_2(14,12,12,10)
54	8^3_1(20,18,16) 8^3_2(20,18,16)
58	8^4_1(20,16,12,10)
62	9^4_1(22,14,14,12)

of edges in each component. 4^2_1 and 8^2_2 are illustrated in figures 6 and 7. Thus, we can conclude that symmetry is not a geometric property of the link, but a topological property.

The minimal edge numbers of some links with more than 2 components were also computed, and the results are in Table II.

The links 6^3_1, 6^3_3, 7^3_1, 8^4_1, 8^4_2, 8^4_3 and 9^4_1 have components which are either linked as 2^2_1 (the Hopf-link), or unlinked. Thus, by theorem 2.1, we can compute upper bounds of the branched edge index for the Hopf-link, $\mu^B_{2^2_1}$. The best upper bound is given by 8^4_3, which is $13\frac{1}{3}$, as illustrated in figure 8.

We next turn our attention briefly to chainmail. In figures 9 and 10 we illustrate two examples of chainmail which we constructed. We conjecture that these are the minimal conformations of these examples (but note that the chainmail in figure 9 has many realizations with the same number of edges). The chainmail in figure 9 has coordination number 4, in the sense that each component (except those on the boundary) is linked to four neighbours. If the chainmail is extented (in the obvious way) to infinity in all directions, then the example in figure 9 acquires a translational invariance, and we call it periodic. The chainmail in figure 10 is also

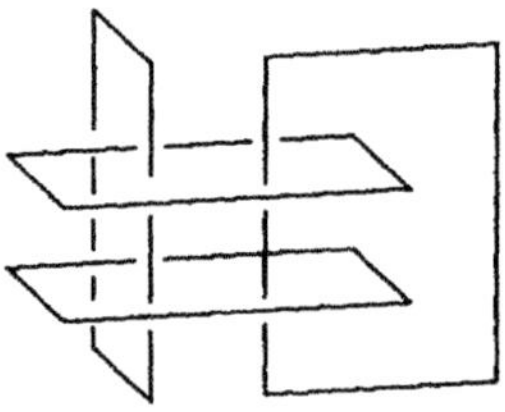

FIG. 8. *The link* 8^4_3 *gives an upper bound of* $13\frac{1}{3}$ *on the branched edge index of the Hopf-link.*

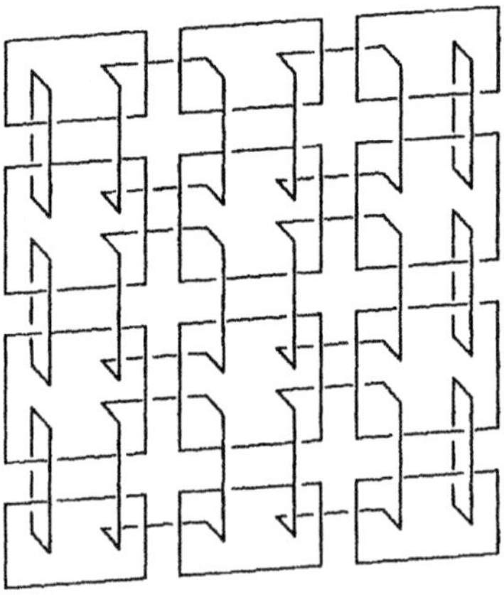

FIG. 9. *Lattice chainmail with coordination number four.*

FIG. 10. *Lattice chainmail with coordination number four.*

periodic. Every pair of components is either not topologically linked, or is the Hopf-link in figures 9 and 10.

It is interesting to observe that the periodic realizations of chainmail induces a partitioning of the plane into tiles with a finite number of shapes, which depends on the angle of projection. In fact, a single periodic realization of a chainmail produces infinite families of tilings of the plane, defined by the angle of projection. The opposite question is more interesting, and harder to answer: Can any tiling be lifted into chainmail?

4. Conclusions. In this paper we have considered the minimal number of edges necessary to tie links of a given type. We proved that the minimal edge number satisfies subadditivity relations, and grows at least linearly with the number of components in the link. In turn, this enabled us to prove existence of the limiting number of edges per component in links (as the number of components is increased without bound). The branched edge index is obtained by assuming that every pair of components in a link is either a link of type L, or the unlink, and it is presumably dependent on the link type L, although we do not have a proof of this fact. We considered also symmetric links, and defined a second edge index for asymmetric links (the bichromatic edge index). The relation to the branched edge index is not known. The mixed branched index of links of type L_1 and L_2 was similarly defined, and shown bounded above by one half the sum of the branched edge indices of the links of type L_1 and L_2 (theorem 2.3).

In section 3 computer estimates of the edge numbers of a variety of links were given. We prove that the number of different links of k or fewer components (k fixed) which can be tied with n edges grows at most exponentially with n.

Acknowledgements: RU was supported by an NSERC (Canada) summer undergraduate award, EJJvR and SGW are supported by NSERC

(Canada) operating grants and EO has support in the form of a fellowship under the EC Human Capital and Mobility Programme.

REFERENCES

[1] C.E. Soteros, D.W. Sumners and S.G. Whittington, *Entanglements and Complexity of Graphs in $\mathbb{Z}^3$*, Math. Proc. Camb. Phil. Soc., **111** (1992) 75–91.

[2] Y. Diao, *Minimal Knotted Polygons on the Cubic Lattice*, J. of Knot. Theory and its Ram., **2** (1993) 413–425.

[3] Y. Diao, *The Number of Smallest Knots on the Cubic Lattice*, J. Stat. Phys., **74** (1994) 1247–1254.

[4] J.W. Alexander and G.B. Briggs, *On Types of Knotted Curves*, Ann. of Math. (2), **28** (1927) 562–586.

[5] E.J. Janse van Rensburg and S.D. Promislow, *Minimal Knots in the Cubic Lattice*, J. of Knot. Theory and its Ram., **4** 115–130.

[6] J.M. Hammersley, *The Number of Polygons on a Lattice*, Proc. Camb. Phil. Soc., **57** (1961) 516–523.

[7] E. Hille, *Functional Analysis and Semi-Groups*, AMS Colloq. Publ., **31** (1948) (New York: AMS).

[8] J.M. Hammersley and K.W. Morton, *Poor Man's Monte Carlo*, J. Roy. Stat. Soc. B, **16** (1954) 23–38.

[9] J.B. Wilker and S.G. Whittington, *Extension of a Theorem on Super-Multiplicative Functions*, J. Phys. A: Math. Gen., **12** (1979) L245–L247.

[10] S. Kirkpatrick, C.D. Gelatt Jr. and M.P. Vecchi, *Optimisation by Simulated Annealing*, Science, **220**, no. 4598 (1983) 671–680.

[11] B. Berg and D. Foester, *Random Paths and Random Surfaces on a Digital Computer*, Phys. Lett., **106B** (1981) 323–326.
C. Aragao de Carvalho, S. Caracciolo and J. Frölich, *Polymers and $g|\phi|^4$ Theory in Four Dimensions*, Nucl. Phys. B [FS7], **215** (1983) 209–248.

[12] E.J. Janse van Rensburg and S.G. Whittington, *The BFACF Algorithm and Knotted Polygons*, J. Phys. A: Math. Gen., **24** (1991) 5553–5567.

[13] J.M. Hammersley, *The Number of Polygons on a Lattice*, Proc. Camb. Phil. Soc., **57** (1961) 516–523.

[14] N. Madras and G. Slade, *The Self-Avoiding Walk*, (Birkhäuser, Boston, 1993).

[15] E.J. Janse van Rensburg, *Lattice Invariants for Knots*, In *Mathematical Approaches to Biomolecular Structure and Dynamics*, Ed J.P. Merirov, K. Schulten and D.W. Sumners, 11–20, Proceedings of the IMA Summer Program in Molecular Biology, July 1994 (Springer-Verlag, New York, 1995).

KNOTS IN GRAPHS IN SUBSETS OF Z^3

C.E. SOTEROS*

Abstract. The probability that an embedding of a graph in Z^3 is knotted is investigated. For any given graph (embeddable in Z^3) without cut edges, it is shown that this probability approaches 1 at an exponential rate as the number of edges in the embedding goes to infinity. Furthermore, at least for a subset of these graphs, the rate at which the probability approaches 1 does not depend on the particular graph being embedded. Results analogous to these are proved to be true for embeddings of graphs in a subset of Z^3 bounded by two parallel planes (a slab).

In order to investigate the knotting probability of embeddings of graphs in a rectangular prism (an infinitely long rectangular tube in Z^3), a pattern theorem for self-avoiding polygons in a prism is proved. From this it is possible to prove that for any given eulerian graph, the probability that an embedding of the graph in a prism is knotted goes to 1 as the number of edges in the embedding goes to infinity. Then, just as for Z^3, there is at least a subset of these graphs for which the rate that this probability approaches 1 does not depend on the particular graph.

Similar results are shown to hold in cases where restrictions are placed on the number of edges per branch in a graph embedding.

Key words. knots, graph embeddings, branched polymer, simple cubic lattice.

AMS(MOS) subject classifications. 82B41,57M25,05C10,05C80,05C30.

1. Introduction. In 1988 Sumners and Whittington [1] investigated questions about knottedness of a closed curve of given length embedded in the three dimensional integer lattice, Z^3. They and, independently, Pippenger [2] showed that sufficiently long closed curves embedded in Z^3 are almost surely knotted. Soteros, Sumners and Whittington [3] extended the results of Sumners and Whittington in several directions. In particular, they showed that all but exponentially few sufficiently long closed curves embedded in Z^3 are highly composite knots, with any knot type appearing numerous times in the prime knot factorization. They also found that similar results hold for embeddings of graphs in Z^3.

These results are important to the study of models of polymers in dilute solution. Embeddings of a graph in Z^3 can be used to represent the possible conformations of a polymer with the structure of the given graph. Concerns about knottedness with regard to polymers are motivated by the fact that closed circular DNA has been observed to be knotted and that entanglements in proteins are believed to be relevant to their properties. There are at least two important questions that one would like to address using this model of polymers. How does knottedness depend on the particular structure of the polymer? How does knottedness depend on the geometrical constraints placed on the space in which the polymer is confined? In this paper, these two questions are addressed by studying

* Department of Mathematics, University of Saskatchewan, Saskatoon, SK, Canada S7N 5E6. This work was supported by NSERC of Canada.

the limiting behaviour, as the number of edges of the embedding goes to infinity, of the probability that an embedding of a graph in a subset of Z^3 is knotted. The subsets of Z^3 considered are Z^3 itself, slabs (subsets of Z^3 bounded by two parallel planes), and rectangular prisms (infinitely long rectangular tubes).

As a step towards answering the first question it is shown here, for each of the three lattice subsets, that there is a large class of graphs (including all planar graphs with no vertices of odd degree) for which the probability that an embedding of a graph in the given subset of Z^3 is knotted goes to one as the number of edges in the embedding goes to infinity and that the exponential rate at which the probability approaches one is independent of the specific graph. Thus, at least in terms of this limiting behaviour and this class of structures, knottedness is not very dependent on the particular structure of the polymer and all but exponentially few embeddings of the polymer are knotted. The approach to proving this result, however, depends on the particular subset of Z^3 considered. While the approach is quite similar for both Z^3 and slab geometries, in the case of the rectangular prism a new pattern theorem for self-avoiding polygons is required and proved. Furthermore, in the prism, the exponential growth rate of the number of embeddings of a graph is shown to be dependent on the particular structure of the graph while in Z^3 or in a slab it is not. Because of this, the class of graphs for which we address questions about knottedness in a prism is smaller than for Z^3 or the slab geometries.

The second question posed above is addressed here by studying the probability that an embedding of a simple closed curve, i.e., a self-avoiding polygon, confined to a particular lattice subset is knotted. This probability has been studied previously in [1,4] and the results from these works are reviewed here. It is shown that the probability is dependent on the lattice subset.

The knottedness of embeddings of graphs in Z^3 has been previously investigated in [3,5] and the results from these works are reviewed and generalized further here. The results presented regarding the knottedness of embeddings of graphs in slab and prism geometries are substantially new. In order to discuss the results in all three subsets, many of the established results for self-avoiding walks, self-avoiding polygons, and embeddings of graphs in the three subsets are reviewed.

The paper is organized as follows. In section 2, the terminology associated with using embeddings of graphs as models of branched polymers is reviewed and important results regarding growth constants for graph embeddings are presented. In section 3, questions related to knots in graphs embedded in Z^3 are addressed. In section 4, results regarding growth constants and knottedness for embeddings of graphs in a slab geometry are presented; graphs embedded in a slab geometry are shown to have properties parallel to those of graphs embedded in the whole space, Z^3. In section 5, the rectangular prism geometry is studied and in this case there are substantial differences from the two previous geometries. A pattern theorem for polygons in a prism, needed for the results presented in section 5, is presented in section 6.

2. Embeddings of graphs in Z^3 as models of branched polymers. In this paper the focus is on lattice models of branched polymers and in particular on models using the three dimensional integer lattice, i.e., the simple cubic lattice. The advantages of lattice models of polymers are that the excluded volume property (each monomer exists in a volume of space to the exclusion of any other monomer) is easily incorporated, that the "lattice polymer" has substantial conformational freedom despite the restrictions of the lattice, and perhaps most importantly, that rigorous mathematical analysis of lattice models is possible. It is expected that lattice models of polymers will exhibit qualitatively the same behaviour as "real" polymers. Furthermore, field theoretic arguments suggest that there exist universal quantites, i.e. critical exponents, which will be exactly the same for lattice models and for real polymers. In this paper, we will focus on lattice models of polymers whose branching structure is specified by an abstract graph. In order to introduce appropriate notation the standard lattice models for linear and ring polymers are reviewed first. Graph theory terminology is used to describe these models. In particular, the three dimensional integer lattice is defined to be the infinite graph with vertex set Z^3 and edge set $\{\{u,v\}|u,v \in Z^3, |u-v|=1\}$. Note that Z^3 will be used to represent either, depending on the context, the three dimensional integer lattice or its vertex set. Similarly for V a set of vertices in Z^3, V will be used to represent either the vertex set V or the subraph of Z^3 induced by this vertex set.

Self-avoiding walks are well known as good models of very long linear polymers in equilibrium in dilute solution. An n-step self-avoiding walk on the simple cubic lattice, Z^3, is an alternating sequence of $n+1$ distinct vertices and n directed edges, $u_0, (u_0,u_1), u_1, (u_1,u_2), u_2, ..., u_{n-1}, (u_{n-1}, u_n)$, u_n, such that the vertices $u_i \in Z^3$ for $i = 0, ..., n$, $u_0 = (0,0,0)$, and for each $i = 0, ..., n-1$ the directed edge (u_i, u_{i+1}) joins two nearest neighbour vertices in Z^3 (i.e. u_{i+1} and u_i differ only in one coordinate with the difference being ± 1). c_n is defined to be the number of n-step self-avoiding walks. As a model of a linear polymer composed of n monomers each n-step self-avoiding walk represents a possible polymer conformation and the self-avoidance ensures that there is an excluded volume around each monomer. The quantity $n^{-1}\log c_n$ thus represents the entropy per monomer of this polymer model and as $n \to \infty$ the limiting entropy per monomer has been proved to exist.

LEMMA 2.1 (HAMMERSLEY 1954 [6]). *The following limit exists:*

$$\kappa \equiv \lim_{n\to\infty} n^{-1}\log c_n. \tag{2.1}$$

κ is also known as the *connective constant* for Z^3. From Lemma 2.1 it is clear that

$$\mu \equiv \lim_{n\to\infty} c_n^{1/n} = e^{\kappa}, \tag{2.2}$$

where μ is known as the *growth constant* for self-avoiding walks in Z^3. Further results about c_n can be found in Madras and Slade [7].

Ignoring the orientation of the edges in a self-avoiding walk yields an undirected self-avoiding walk. Let w_n be the number of n-edge undirected self-avoiding walks (up to translation). Clearly,

$$w_n = \frac{c_n}{2} \tag{2.3}$$

and hence the growth constant for w_n will be μ.

Ring polymers in dilute solution have been successfully modelled using self-avoiding polygons. An n-edge *self-avoiding polygon* in Z^3 is an alternating sequence of n distinct vertices and n undirected edges, $u_0, \{u_0, u_1\}, u_1, \{u_1, u_2\}, u_2, ..., u_{n-1}, \{u_{n-1}, u_0\}, u_0$, such that for each $i = 0, ..., n-1$ the vertex $u_i \in Z^3$ and the edge $\{u_i, u_{i+1}\}$ joins two nearest neighbour vertices in Z^3. p_n is defined to be the number of n-vertex self-avoiding polygons (up to translation). Hammersley [8] proved that the growth constant for self-avoiding polygons in Z^3 is the same as that for self-avoiding walks.

LEMMA 2.2 (HAMMERSLEY 1961 [8]).

$$\lim_{n\to\infty} (2n)^{-1} \log p_{2n} = \kappa. \tag{2.4}$$

In order to obtain relationships between other lattice models and self-avoiding walks or polygons, it is also useful to study self-avoiding walks or polygons in wedges [9,10]. The following result is a consequence of the results for self-avoiding polygons in wedges in [10]. Given non-negative integers $\infty \geq \beta > \alpha > 0$, define the wedge $W(\alpha, \beta)$ to be the sublattice of Z^3 induced by the vertex set $\{(x, y, z) \in Z^3 | \alpha x \leq y \leq \beta x + \alpha + 1, x \geq 0, z \geq 0\}$. This definition of the wedge ensures that at least one self-avoiding polygon of arbitrary length contained in the (x, y)-plane can fit in the wedge. Let $p_n^{\alpha,\beta}$ be the number of n-edge self-avoiding polygons confined to $W(\alpha, \beta)$ and containing an edge between $(0, 0, 0)$ and $(0, 1, 0)$; similarly, let $c_n^{\alpha,\beta}$ be the number of n-step self-avoiding walks confined to $W(\alpha, \beta)$ and having its first step from $(0, 0, 0)$ to $(0, 1, 0)$. The wedge arguments in [10] lead to the following:

LEMMA 2.3 (SOTEROS 1992 [10]). *Given non-negative integers α and β such that $\infty \geq \beta > \alpha > 0$,*

$$\lim_{n\to\infty} (2n)^{-1} \log p_{2n}^{\alpha,\beta} = \lim_{n\to\infty} n^{-1} \log c_n^{\alpha,\beta} = \kappa. \tag{2.5}$$

Hence confinement of a self-avoiding polygon or walk to a wedge such as $W(\alpha, \beta)$ does not affect its growth constant. A very general wedge result has been obtained recently by Soteros *et al* [11]. From their results Lemma 2.3 is extended to the case that α and β are any real numbers.

The undirected self-avoiding walk and polygon models are included in a wider class of polymer models in which the structure of a polymer is given by an abstract graph τ and polymer conformations are represented by embeddings of τ in Z^3. In this wider class of models, the graph τ which represents the topology of a polymer is assumed to have no vertices of degree two (except in the case of a ring polymer); the *degree* of a vertex of a graph is defined to be the number of edges incident on the vertex and a *loop* (an edge from a vertex to itself) is assumed to contribute twice to the degree of a vertex. An edge of τ is used to represent a *branch* of the polymer. In the special case of a ring polymer τ is taken to be the *circle graph* which is defined to consist of one vertex (a vertex of degree two) and one loop. The circle graph is thus said to have one branch. Vertices of τ having degree greater than one are referred to as *branch points*. Conformations of a polymer with topology τ are represented by embeddings of τ in Z^3. An *embedding of τ in Z^3* is defined to be any subgraph of Z^3 which is isomorphic to τ when vertices of degree two are suppressed; that is, an embedding of τ is any subgraph of Z^3 which is *homeomorphic* to τ. Define G_r to be the set which includes the circle graph and the set of all connected graphs having at least one edge and no vertices of degree two and whose branch points all have degree less than or equal to r. Soteros *et al* [3] proved that an embedding of τ in Z^3 exists for any $\tau \in G_6$. Given $\tau \in G_6$, define $g_n(\tau)$ to be the number of embeddings of τ in Z^3 (up to translation) consisting of n edges.

Thus an n-edge self-avoiding polygon is an embedding of the circle graph consisting of n edges and $g_n(\text{circle}) = p_n$. Figure 1 b) shows the circle graph along with a 10 edge embedding of the circle graph in Z^3. Define the *line graph* to be the graph consisting of two vertices and one edge which joins the two vertices together. An n-edge undirected self-avoiding walk is thus an n-edge embedding of the line graph and $g_n(\text{line}) = w_n$. Figure 1 a) shows the line graph and a 10 edge embedding of the line graph in Z^3. Define the *figure eight graph* to be the graph with one vertex and two loops. Figure 1 c) depicts a figure eight graph and a 20-edge embedding of it in Z^3.

Based on the following result, the *growth constant*, $\lim_{n\to\infty}(g_n(\tau))^{1/n}$, for the number of embeddings of a graph exists and is independent of the specific graph.

LEMMA 2.4 (SOTEROS, SUMNERS AND WHITTINGTON 1992 [3]). *Given a graph $\tau \in G_6$*

$$\lim_{n\to\infty} n^{-1} \log g_n(\tau) = \kappa \tag{2.6}$$

where κ is the connective constant for self-avoiding walks and where if τ admits no embeddings with an odd number of edges then the limit is taken only through even values of n .

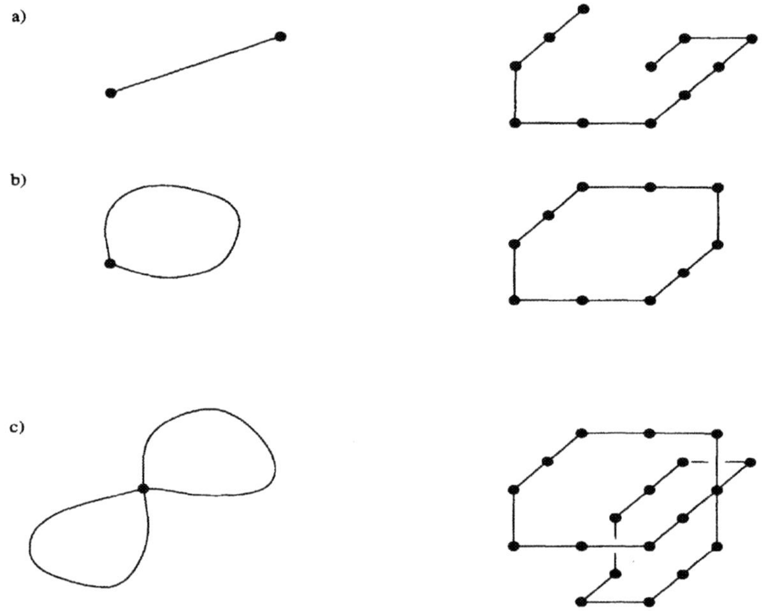

FIG. 1. *In a) the line graph and a 10-edge embedding of the line graph in Z^3 is depicted. In b) the circle graph and a 10-edge embedding of the circle graph in Z^3 is depicted. In c) the figure eight graph and a 20-edge embedding of the figure eight graph in Z^3 is depicted.*

Proof. The proof of this Lemma is reviewed here in order to introduce a basic approach to obtaining results concerning growth constants of graph embeddings. Let τ be any graph in G_6 and fix a labelling of the vertices and edges of τ. Let f be the number of edges in τ. We first note that given any embedding of τ, the vertices (except for those of degree two) and branches of the embedding can be labelled according to the labelling given for τ. In particular, label the vertices first and, when ambiguity arises in the labelling of vertices, label them according to the lexicographic ordering of their coordinates in Z^3. When ambiguity arises in the labelling of branches (i.e. two or more branches have the same labelled vertices at their ends), label them according to the lexicographic ordering of the coordinates of the vertex of each branch that is adjacent to its smallest end vertex.

An upper bound is obtained for $g_n(\tau)$ in terms of self-avoiding walks as follows. Consider any n-edge embedding of τ. Translate the embedding so that the first branch has its smallest (in terms of the labelling just described) end vertex at the origin. For each branch, orient the edges of the branch to go from its smaller to its larger end vertex. Thus each branch of the embedding can be viewed as a self-avoiding walk and therefore

$$(2.7) \qquad g_n(\tau) \leq \sum_{\{m_i | \sum_{i=1}^{f} m_i = n\}} c_{m_1} c_{m_2} \cdots c_{m_f} \leq e^{\kappa n + o(n)}$$

where m_i represents the number of edges in branch i of τ.

A lower bound is obtained for $g_n(\tau)$ in terms of self-avoiding polygons as follows. Start with any embedding of τ in which the rightmost (maximum x-coordinate) plane of the embedding, $x = k$ say, contains the edge $\{(k, 0, 0), (k, 1, 0)\}$. It is always possible to find such an embedding with an even number of edges and, if there exists an embedding of τ with an odd number of edges, it is also possible to find such an embedding with an odd number of edges [3]. Let $m_e > 0$ be an even number for which such an m_e-edge embedding of τ exists and m_o be an odd number for which such an m_o-edge embedding of τ exists. For any $2n \geq m_e + 4$, let P be a $(2n - m_e)$-edge self-avoiding polygon confined to $W(0, \infty)$ and containing the edge $\{(0, 0, 0), (0, 1, 0)\}$. Translate P so that $(0, 0, 0)$ is translated to $(k + 1, 0, 0)$ and then *concatenate* P to the embedding of τ by adding the edges $\{(k, 1, 0), (k + 1, 1, 0)\}$ and $\{(k, 0, 0), (k + 1, 0, 0)\}$ and then deleting the edges $\{(k, 0, 0), (k, 1, 0)\}$ and $\{(k+1, 0, 0), (k+1, 1, 0)\}$. This results in the upper bound

$$(2.8) \qquad p^{0,\infty}_{2n-m_e} \leq g_{2n}(\tau).$$

If τ has embeddings with an odd numbers of edges, a similar concatenation argument yields

$$(2.9) \qquad p^{0,\infty}_{2n+1-m_o} \leq g_{2n+1}(\tau).$$

If τ does not admit any odd edge embeddings (for example if τ is the circle or the figure eight graph (see Figure 1 c)), taking logarithms, dividing by $2n$, letting $n \to \infty$, and using Lemma 2.3 in equations (2.7) (with n replaced by $2n$) and (2.8) gives the required result where the limit is taken only through even values. If τ does admit odd edge embeddings, equations (2.7), (2.8), and (2.9) lead to the required result where the limit is taken through all values of n. □

Polymer chemists are frequently interested in the case that the branches of a polymer all consist of the same number of monomers (i.e. the polymer chains making up the branched polymer structure are monodisperse). Thus it is useful to consider the case where the distribution of edges to the branches of a graph embedding is fixed. Given $\tau \in G_6$, let f be the number of edges of τ. Suppose $f \geq 2$. Given a fixed labelling of the vertices and edges of τ, we can define $g_n(\tau; (\phi_1, \phi_2, ..., \phi_f))$ to be the number of embeddings of τ in Z^3, distinct up to translation, composed of n edges with ϕ_i edges in its ith branch. The ϕ_i are positive integers such that $\sum_i \phi_i = n$ and the labelling of the branches of the embedding is determined, as described in the proof of Lemma 2.4, by the fixed labelling of τ. The ordered set of numbers $\phi \equiv (\phi_1, \phi_2, ..., \phi_f)$ will be referred to as the *edge distribution* of the embedding. In [10], the number of *uniform* or monodisperse embeddings of a graph was studied, $g_{nf}(\tau; (n, n, ..., n))$, and the following lemma resulted.

LEMMA 2.5 (SOTEROS 1992 [10]). *Given a bipartite graph $\tau \in G_6$*

$$\lim_{n\to\infty} (nf)^{-1} \log g_{nf}(\tau; (n, n, ..., n)) = \kappa. \tag{2.10}$$

Given a graph $\tau \in G_6$ which is not bipartite (i.e. has at least one cycle of odd length)

$$\lim_{n\to\infty} (2nf)^{-1} \log g_{2nf}(\tau; (2n, 2n, ..., 2n)) = \kappa. \tag{2.11}$$

Proof. The basic idea of the proof is to first get an upper bound on $g_{nf}(\tau; (n, n, ..., n))$ in terms of self-avoiding walks by noting that $g_{nf}(\tau; (n, n, ..., n)) \leq g_{nf}(\tau)$ and then using equation (2.7). A lower bound for $g_{nf}(\tau; (n, n, ..., n))$ is obtained by starting with an embedding of τ such that:

(i) Exactly one edge of each branch of τ lies in the rightmost plane, $x = k$, of the embedding. These rightmost edges lie in the line $z = 0$, $x = k$, $y \geq 0$.
(ii) The parity (even or odd) of the number of edges in a branch is the same for all branches.
(iii) Each edge in the line $z = 0$, $x = k$ is at least f edges apart from any other edge in the line.

Then, in order to create a new embedding of τ, for each $i = 1, ..., f$, a polygon in $W(i-1, i)$ is concatenated to the edge of the ith branch of τ contained in the rightmost plane $x = k$. This gives the lower bound

$$p^{0,1}_{n-m_1} p^{1,2}_{n-m_2} \cdots p^{f-1,f}_{n-m_f} \leq g_{nf}(\tau; (n, n, ..., n)) \tag{2.12}$$

where $(m_1, m_2, ..., m_f)$ is the edge distribution of the initial embedding of τ. Taking logarithms, dividing by nf, letting n go to infinity, and using Lemma 2.3 gives the required result. □

This type of argument can also be used to obtain results about other types of edge distributions. For example, consider the sequence of numbers $g_N(\tau; (\phi_1(n), ..., \phi_f(n)))$ for $n = 1, 2, 3, ...$, where $\phi_i(n)$ is a postive integer valued function of n and $N = \sum_{i=1}^{f} \phi_i(n)$. Let $\phi(n) \equiv (\phi_1(n), ..., \phi_f(n))$. In order for $g_N(\tau; \phi(n)) > 0$ when τ contains a cycle, constraints must be placed on the parity of the components of $\phi(n)$ to ensure that any cycle in an embedding of τ in Z^3 consists of an even number of edges. In particular, each embedding of a cycle must contain an even number of odd length branches. Define the *parity distribution* associated with an edge distribution $\phi(n)$ to be the f-tuple of parities whose ith component is the parity of $\phi_i(n)$. Suppose that $g_N(\tau; \phi(n)) > 0$ for $n \geq 1$ and, without loss of generality, that the sequence $\{\phi(n)\}_{n\geq 1}$ has the property that given an

$i \in \{1, 2, ..., f\}$ the parity of $\phi_i(n)$ is fixed for all $n \geq 1$, i.e., the sequence of $\phi(n)$'s has a fixed parity distribution. If the sequence does not have a fixed parity distribution, one considers separately each subsequence of $\{\phi(n)\}_{n\geq 1}$ that does have a fixed parity distribution; there can be at most 2^f such subsequences. Further suppose that given an $i \in \{1, 2, ..., f\}$ either $\phi_i(n)$ is a constant function or a strictly increasing function of n. If $\phi_i(n)$ is strictly increasing, we refer to the ith branch as *growing*; otherwise, the ith branch is referred to as *constant*. An upper bound for $g_N(\tau; \phi(n))$ is obtained in the same manner as in the proof of Lemma 2.4, namely, each embedding of τ is separated into f self-avoiding walks. To get a lower bound for $g_N(\tau; \phi(n))$ we again need to construct an appropriate embedding to which we can concatenate polygons in wedges. Since embeddings of τ exist with the parity and edge distributions given by $\{\phi(n)\}_{n\geq 1}$, it should be possible to find an embedding of τ having $\phi(1)$'s parity distribution, having $\phi_i(1)$ edges in each constant branch i, and having enough space between each branch of the embedding so that for each growing branch it is possible to concatenate a polygon to an edge of the branch and extend the branch so that it has a unique edge in the rightmost plane of the embedding. In this way one can construct an embedding of τ with properties (i) and (iii) as in the proof of Lemma 2.5 and with property (ii) revised as follows:

(ii*) The number of edges in branch i has the same parity as $\phi_i(1)$ and equals $\phi_i(1)$ if branch i is constant.

Establishing an appropriate lower bound then follows by concatenating polygons in wedges to the edges in the line $z = 0$, $x = k$ just as in the proof of Lemma 2.5. Thus, we obtain the following result.

LEMMA 2.6. *Given $\tau \in G_6$ with f edges, let $\{\phi(n)\}_{n\geq 1}$ be a sequence of edge distributions such that for $1 \leq i \leq f$, $\phi_i(n)$ is either a constant or strictly increasing function of n. If for each subsequence of $\{\phi(n)\}_{n\geq 1}$ with fixed parity distribution an embedding of τ with the properties (i), (ii*), and (iii) exists, then*

$$\lim_{n\to\infty} N^{-1} \log g_N(\tau; \phi(n)) = \kappa, \tag{2.13}$$

where $N = \sum_{i=1}^{f} \phi_i(n)$.

Thus the growth constant for embeddings of a graph does not depend on the fixed edge distribution sequence given for the graph.

3. Knots in graphs in Z^3. In this section, results about knots in graph embeddings in Z^3 are discussed. Unless stated otherwise, the terminology used here will be the same as that defined in Soteros, Sumners and Whittington [3]. Any locally flat piecewise linear embedding of τ in 3-space is said to be a knot of τ. The set of equivalence classes of knots of τ (as defined in [3]) is denoted by $\mathcal{K}(\tau)$. For $K \in \mathcal{K}(\tau)$, the crossing number of K, $C(K)$, is the minimum number of crossings obtainable in any knot diagram for any knot $k \in K$ and we define

$$N_{\min}(\tau) = \min_{K \in \mathcal{K}(\tau)} C(K). \tag{3.1}$$

The knottedness of embeddings of a graph will be discussed in terms of embeddings in 3-space. Thus, in order to establish a point of reference, define $K \in \mathcal{K}(\tau)$ to be *unknotted* if there are members of K which realize $N_{\min}(\tau)$; in this case all members of K are said to be unknotted. Otherwise, $K \in \mathcal{K}(\tau)$ and all members of K are said to be *knotted*. An embedding of τ in Z^3 will be considered knotted if as an embedding in 3-space it is knotted.

Note that any embedding of the graph K_7 in 3-space contains a knotted cycle [12]. Although K_7 is not embeddable in Z^3, this example suggests the possibity that, based on the definition above, an unknotted embedding of a graph in Z^3 may contain a cycle which is a knotted self-avoiding polygon. It will be useful later to consider the set of graphs embeddable in Z^3 for which this is not possible. Thus, for $2 \le r \le 6$, let $\tilde{G}_r \subseteq G_r$ be the set of graphs such that for $\tau \in \tilde{G}_r$ any cycle in an unknotted embedding of τ in Z^3 is an unknotted embedding of a circle graph (i.e., an unknotted self-avoiding polygon). Note that $\tilde{G}_6$ contains the set of all planar graphs in G_6.

Given Lemmas 2.1-2.6, the main additional ingredient that is needed in order to prove results about knotted embeddings of a graph τ in Z^3 is an appropriate pattern theorem. A *pattern* will be defined to be any embedding of the line graph in Z^3. A *Kesten* pattern will be defined to be any pattern which appears at least three times in an embedding of the line graph. Let $g_n(\tau; \bar{P})$ be the number of n-edge embeddings of τ in Z^3 which do not contain the pattern P; let $g_n(\tau; P)$ be the number of n-edge embeddings of τ in Z^3 which contains the pattern P. Similarly, let $g_N(\tau; \phi(n), \bar{P})$ be the number of embeddings of τ in Z^3 composed of $N = \sum_{i=1}^{f} \phi_i(n)$ edges, with edge distribution $\phi(n) = (\phi_1(n), \phi_2(n), ..., \phi_f(n))$, and such that the pattern P does not appear in the embedding. A pattern theorem due to Kesten for self-avoiding walks yields the following pattern theorem for embeddings of the line graph.

LEMMA 3.1 (KESTEN 1963 [13]). *Let P be any Kesten pattern.*

$$\lim_{n \to \infty} n^{-1} \log g_n(\text{line}; \bar{P}) = \kappa(\bar{P}) < \kappa. \tag{3.2}$$

Combining this with Lemma 2.1 gives that

$$\lim_{n \to \infty} \frac{g_n(\text{line}; P)}{g_n(\text{line})} = \lim_{n \to \infty} 1 - e^{(\kappa(\bar{P}) - \kappa)n} = 1. \tag{3.3}$$

Hence, for n sufficiently large, all but exponentially few embeddings of the line graph contain the pattern P and

$$\lim_{n \to \infty} n^{-1} \log g_n(\text{line}; P) = \kappa. \tag{3.4}$$

Based on equation (2.7), one can obtain an upper bound for $g_n(\tau)$ in terms of $g_n(\text{line})$ by using the fact that $c_n = 2g_n(\text{line})$. The same reasoning allows one to obtain an upper bound for $g_n(\tau;\bar{P})$ in terms of $g_n(\text{line};\bar{P})$ and for $g_N(\tau;\phi(n),\bar{P})$ in terms of $g_n(\text{line};\bar{P})$ and gives the following two results.

LEMMA 3.2 (SOTEROS, SUMNERS AND WHITTINGTON 1992 [3]). *Let P be any Kesten pattern and τ any graph in G_6.*

$$\limsup_{n\to\infty} n^{-1}\log g_n(\tau;\bar{P}) \le \kappa(\bar{P}) < \kappa. \tag{3.5}$$

That is, for sufficiently large n, the pattern P appears at least once in all but exponentially few embeddings of τ.

LEMMA 3.3. *Let P be any Kesten pattern and τ any graph in G_6. Given any edge distribution sequence $\{\phi(n)\}_{n\ge1}$ for which equation (2.13) holds with $N = \sum_{i=1}^{f}\phi_i(n)$,*

$$\limsup_{n\to\infty} N^{-1}\log g_N(\tau;\phi(n),\bar{P}) \le \kappa(\bar{P}) < \kappa. \tag{3.6}$$

That is, for sufficiently large n, the pattern P appears at least once in all but exponentially few embeddings of τ when the edge distribution is fixed.

A version of Lemma 3.3 for uniform embeddings of τ was given in [5].

In order to discuss knots in embeddings of τ in Z^3, the focus is on the pattern T^q which is formed by concatenating a sequence of q tight trefoils, T, where $q = \lceil 1 + \frac{N_{\min}(\tau)}{3}\rceil$. Figure 2 depicts a *tight trefoil*, T; if T appears in a self-avoiding polygon the polygon must be knotted. If T^q appears in a cycle in an embedding of τ in Z^3, then the crossing number of the embedding is $\ge 3q \ge N_{\min}(\tau)$ [3] and hence the embedding is knotted.

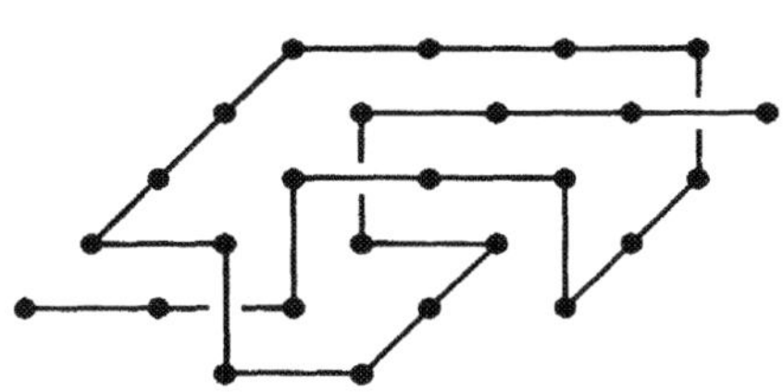

FIG. 2. *A tight trefoil pattern T in Z^3.*

Define $g_n^o(\tau)$ to be the number of unknotted n-edge embeddings of τ. Clearly if τ has no cycles then $g_n^o(\tau) = g_n(\tau)$. Sumners and Whittington [1] used a concatenation argument to show that for $p_n^o \equiv g_n^o(\text{circle})$

$$p_m^o p_n^o \le p_{m+n}^o \tag{3.7}$$

and then they used Kesten's pattern theorem to prove the following result for unknotted self-avoiding polygons.

LEMMA 3.4 (SUMNERS AND WHITTINGTON 1988 [1]).

$$\lim_{n\to\infty}(2n)^{-1}\log p_{2n}^o \equiv \kappa_o < \kappa \tag{3.8}$$

and hence the probability that an n-edge self-avoiding polygon is knotted, $\frac{p_n - p_n^o}{p_n}$*, goes to unity as*

$$1 - e^{-\alpha n + o(n)} \tag{3.9}$$

when $n \to \infty$*, with* $\alpha \equiv \kappa - \kappa_o$.

The following theorem about the knottedness of graph embeddings in Z^3 is a consequence of Lemmas 3.2 and 3.4. Only graphs in which each branch of the graph is contained in a cycle need be considered; that is, graphs which have no cut edges. (A *cut edge* is any edge which when removed disconnects the graph.) An important subset of the graphs with no cut edges is the set of eulerian graphs; a graph is said to be *eulerian* if there exists a closed walk on the graph which uses each edge of the graph exactly once. Note that τ is eulerian if and only if every vertex of τ has even degree. The θ-graph (the graph with two vertices of degree 3 and three edges, each going from one vertex of the graph to the other) is an example of a graph with no cut edges that is not eulerian. The circle graph and the figure eight graph are eulerian.

THEOREM 3.1. *Suppose* $\tau \in G_6$ *has no cut edges,*

$$\kappa_o \leq \liminf_n n^{-1}\log g_n^o(\tau) \leq \limsup_{n\to\infty} n^{-1}\log g_n^o(\tau) \leq \kappa(\bar{T^q}) < \kappa \tag{3.10}$$

(where as in Lemma 2.4 the limit is taken only through values of n for which $g_n^o(\tau) > 0$*) and hence the probability that an n-edge embedding of* τ *is knotted goes to unity at least as fast as*

$$1 - e^{-\beta n + o(n)} \tag{3.11}$$

and no faster than

$$1 - e^{-\alpha n + o(n)} \tag{3.12}$$

when $n \to \infty$*, with* $\beta = \kappa - \kappa(\bar{T^q})$ *and* α *as defined in equation (3.9).*

If τ *is eulerian and* $\tau \in \tilde{G}_6$*,*

$$\lim_{n\to\infty}(2n)^{-1}\log g_{2n}^o(\tau) = \kappa_o \tag{3.13}$$

and hence the probability that an n-edge embedding of τ *is knotted goes to unity as*

$$1 - e^{-\alpha n + o(n)} \tag{3.14}$$

when $n \to \infty$.

Proof. For any $\tau \in G_6$, starting with an unknotted embedding of τ and then concatenating an unknotted embedding of a polygon yields a new unknotted embedding of τ. Thus following an argument similar to the one that leads to equations (2.8) and (2.9), there exists m such that

$$p_{n-m}^{o,0,\infty} \leq g_n^o(\tau) \tag{3.15}$$

where $p_n^{o,0,\infty}$ is the number of n-edge unknotted self-avoiding polygons in the wedge $W(0,\infty)$. Here we note that, given Lemma 3.4, Lemma 2.3 can also be proved for unknotted polygons in wedges with κ replaced by κ_o. Thus taking logarithms, dividing by n, and letting n go to infinity in equation (3.15) gives

$$\kappa_o \leq \liminf_n n^{-1} \log g_n^o(\tau) \tag{3.16}$$

where if τ does not admit any odd edge embeddings then the limit is taken only through even values of n.

In the case $\tau \in G_6$ has no cut edges, an upper bound on $g_n^o(\tau)$ can be obtained as in equation (2.7) except now no branch of τ can contain the pattern T^q. Thus

$$\begin{aligned} g_n^o(\tau) \leq \quad & 2^f \sum_{\{m_i \mid \sum_{i=1}^f m_i = n\}} g_{m_1}(\text{line}, \bar{T^q}) g_{m_2}(\text{line}, \bar{T^q}) \cdots g_{m_f}(\text{line}, \bar{T^q}) \\ (3.17) \leq \quad & e^{\kappa(\bar{T^q})n + o(n)}. \end{aligned}$$

Equation (3.17) and equation (3.16) lead to equation (3.10). (Equation (3.17) was first presented in [3].) The limiting behaviour of the probability that an n-edge embedding of τ is knotted is determined by equation (3.10) since it implies that for sufficiently large n

$$1 - e^{-\alpha n + o(n)} \geq \frac{g_n(\tau) - g_n^o(\tau)}{g_n(\tau)} \geq 1 - e^{-\beta n + o(n)} \tag{3.18}$$

with α and β as defined in the statement of the theorem.

If τ is eulerian then there exists a closed path which uses every edge of τ exactly once. Fix such a closed path. This path can be used to decompose an embedding of τ into at most f rooted (or possibly multiply rooted) self-avoiding polygons. Let f_p be the number of polygons that τ decomposes into. For each vertex of degree $2j$ of τ, j roots are needed. Thus $2n_4 + 3n_6$ roots must be distributed to the f_p polygons, where n_i is the number of vertices of degree i in τ. An upper bound on the number of ways to root and label the f_p polygons is given by

$$R_\tau(n) = 2^{f_p} \binom{f_p + 2n_4 + 3n_6 - 1}{2n_4 + 3n_6} n^{(2n_4+3n_6)f_p} \tag{3.19}$$

where n is the total number of edges in an embedding of τ. Recall that if $\tau \in \tilde{G}_6$, every cycle in an unknotted embedding of τ is unknotted. Thus for eulerian $\tau \in \tilde{G}_6$ the following upper bound for $g^o_{2n}(\tau)$ results

$$(3.20)\; g^o_{2n}(\tau) \leq R_\tau(2n) \sum_{\{m_i | \sum_{i=1}^{f_p} m_i = n\}} p^o_{2m_1} p^o_{2m_2} \cdots p^o_{2m_{f_p}} \leq e^{2n\kappa_o + o(n)}$$

where we have used equation (3.7) and Lemma 3.4 for the rightmost inequality. This combined with equation (3.16) gives equation (3.13). □

For any $\tau \in G_6$ let $\{\phi(n)\}_{n\geq 1}$ be an edge distribution sequence for which equation (2.13) holds. We define a *fast growing* branch of τ to be a branch i for which

$$1 \geq \lim_{n\to\infty} \frac{\phi_i(n)}{N} > 0. \tag{3.21}$$

Using this definition we obtain the following result.

THEOREM 3.2. *Given $\tau \in G_6$ with f edges and at least one cycle, if $\{\phi(n)\}_{n\geq 1}$ is an edge distribution sequence such that Lemma 2.6 holds, such that there exists an unknotted embedding of τ with edge distribution $\phi(1)$, and such that at least one branch of a cycle of τ is fast growing, then*

$$\begin{aligned} \kappa_o \leq\ & \liminf_{n\to\infty} N^{-1} \log g^o_N(\tau;\phi(n)) \\ \leq\ & \limsup_{n\to\infty} N^{-1} \log g^o_N(\tau;\phi(n)) \leq \kappa(\bar{T}^q) < \kappa. \end{aligned} \tag{3.22}$$

Furthermore, if all the fast growing branches of $\tau \in \tilde{G}_6$ are contained in an eulerian subgraph (not necessarily connected) of τ

$$\lim_{n\to\infty} N^{-1} \log g^o_N(\tau;\phi(n)) = \kappa_o. \tag{3.23}$$

Proof. Let τ be any fixed graph in G_6 and let $\{\phi(n)\}_{n\geq 1}$ be an edge distribution sequence for which Lemma 2.6 holds. Based on the argument that leads to equation (2.12), concatenating unknotted polygons in wedges to an unknotted embedding of τ with properties (i), (ii*), and (iii) results in the lower bound

$$p^{o,0,1}_{\phi_1(n)-m_1} p^{o,1,2}_{\phi_2(n)-m_2} \cdots p^{o,f-1,f}_{\phi_f(n)-m_f} \leq g^o_N(\tau;\phi(n)) \tag{3.24}$$

where $(m_1, m_2, ..., m_f)$ is the edge distribution of the initial unknotted embedding of τ. Taking logarithms, dividing by N, and letting n go to infinity then gives

$$\kappa_o \leq \liminf_{n\to\infty} N^{-1} \log g^o_N(\tau;\phi(n)). \tag{3.25}$$

Let i be a fast growing branch contained in a cycle of τ. To obtain the upper bound, we separate an n-edge embedding of τ into f embeddings of the line graph (as in equation (3.17)) except that the ith branch cannot contain the pattern T^q. Thus

$$\begin{aligned} g_N^o(\tau;\phi(n)) &\leq 2^f w_{\phi_1(n)}\cdots w_{\phi_{i-1}(n)} w_{\phi_{i+1}(n)}\cdots w_{\phi_f(n)} g_{\phi_i(n)}(\text{line},\bar{T^q}) \\ &\leq e^{\kappa(N-\phi_i(n))+\kappa(\bar{T^q})\phi_i(n)+o(N)}. \end{aligned} \tag{3.26}$$

Taking logarithms, dividing by N, and letting n go to infinity in this equation gives

$$\limsup_{n\to\infty} N^{-1}\log g_N^o(\tau;\phi(n)) < \kappa. \tag{3.27}$$

Further suppose that all fast growing branches of $\tau \in \tilde{G}_6$ are contained in an eulerian subgraph η of τ. Let f_p be the number of polygons that η separates into and let $R_\eta(n)$ be as defined in equation (3.19). Assume that the remaining branches of τ (i.e. those not part of η) are labelled from 1,2,...,f'. Starting with an N-edge unknotted embedding of τ with edge distribution $\phi(n)$, none of the polygons that η separates into can be knotted hence,

$$\begin{aligned} g_N^o(\tau;\phi(n)) &\leq R_\eta(N)2^{f'} w_{\phi_1(n)}\cdots w_{\phi_{f'}(n)} p_{m_1}^o\cdots p_{m_{f_p}}^o \\ &\leq e^{\kappa_o N+o(N)} \end{aligned} \tag{3.28}$$

where m_i is the number of edges of the embedding in η's ith polygon and the fact that $\lim_{n\to\infty}\frac{\sum_{i=1}^{f'}\phi_i(n)}{N} = 0$ has been used to obtain the rightmost inequality. Equations (3.25) and (3.28) lead to equation (3.23). □

Note that this theorem was proved for the uniform case (i.e., $\phi(n) = (n,n,...,n)$) in [5]; a related theorem was proved in [3] for embeddings in which the fraction of edges in a designated cycle was fixed rather than the whole edge distribution being fixed. One can also use a more general version of Kesten's pattern theorem to prove that in the cases covered by Theorems 3.1 and 3.2 almost all the considered embeddings of a graph τ are highly composite knots [3,5].

In the next sections the extent to which these results hold for embeddings of graphs in subsets of Z^3 are explored.

4. Knots in graphs in an L-slab. An L-slab is the sublattice of Z^3 induced by the vertex set $\{(x,y,z) \in Z^3 | 0 \leq z \leq L\}$. The properties of self-avoiding walks and polygons in an L-slab have been reviewed in two recent papers by Tesi *et al* [4,14]. In order to address questions about knots in graphs in an L-slab, the same ingredients as those introduced in the last two sections are needed, namely, a result equating the growth constant for self-avoiding walks and self-avoiding polygons (as in Lemma 2.2), a result for the growth constant of self-avoiding polygons in wedges (as in Lemma 2.3), and a pattern theorem for self-avoiding walks (as in Lemma 3.1). In fact the arguments which led to the results just listed can be modified in a straightforward fashion to deal with self-avoiding polygons and walks in an L-slab, $L \geq 1$.

Define $p_n(L)$ to be the number (up to translation) of n-edge self-avoiding polygons confined to an L-slab and let $c_n(L)$ to be the number of n-step self-avoiding walks confined to an L-slab. From the work of Hammersley and Whittington (1985) [9] and Madras and Slade (1993) [7] we have that

LEMMA 4.1.

$$\lim_{n\to\infty} n^{-1}\log c_n(L) = \lim_{n\to\infty}(2n)^{-1}\log p_{2n}(L) = \kappa(L) < \kappa \tag{4.1}$$

where $\kappa(L)$ is strictly increasing in L and $\lim_{L\to\infty}\kappa(L) = \kappa$.

Given non-negative integers $\infty \geq \beta > \alpha > 0$, define $W(\alpha,\beta,L)$ to be the sublattice of an L-slab induced by the vertex set $\{(x,y,z) \in Z^3 | \alpha x \leq y \leq \beta x + \alpha + 1, 0 \leq x, 0 \leq z \leq L\}$. For $L \geq 0$, let $p_n^{\alpha,\beta}(L)$ be the number of n-edge self-avoiding polygons confined to $W(\alpha,\beta,L)$ and containing an edge between $(0,0,0)$ and $(0,1,0)$; similarly, let $c_n^{\alpha,\beta}(L)$ be the number of n-step self-avoiding walks confined to $W(\alpha,\beta,L)$ and having its first step from $(0,0,0)$ to $(0,1,0)$. The wedge arguments in [10,11] are easily modified to give:

LEMMA 4.2.

$$\lim_{n\to\infty}(2n)^{-1}\log p_{2n}^{\alpha,\beta}(L) = \lim_{n\to\infty} n^{-1}\log c_n^{\alpha,\beta}(L) = \kappa(L). \tag{4.2}$$

Tesi *et al* [4,14] have proved an appropriate pattern theorem for self-avoiding walks in an L-slab. Let $C_{m_1,m_2}(L)$ be the set of self-avoiding walks completely contained in $D_{m_1,m_2}(L) = \{(x,y,z) \in Z^3 | 0 \leq x \leq m_1, 0 \leq y \leq m_2, 0 \leq z \leq L\}$ with one endpoint at the origin and the other at $(m_1, m_2, 0)$. For any $m_1 > 0$ and $m_2 > 0$, a $K_{m_1,m_2}(L)$ pattern is defined to be an self-avoiding walk which appears in some element of $C_{m_1,m_2}(L)$. The following pattern theorem holds.

LEMMA 4.3 (TESI *et al* [4,14]). *For any $m_1 > 0$ and $m_2 > 0$ and any $K_{m_1,m_2}(L)$ pattern P,*

$$\lim_{n\to\infty} n^{-1}\log c_n(L,\bar{P}) \equiv \kappa(L,\bar{P}) < \kappa(L) \tag{4.3}$$

where $c_n(L,\bar{P})$ is the number of n-step walks in an L-slab which do not contain the pattern P.

Arguments such as those given in [3] can be used to show that, for $L \geq 2$ and any graph $\tau \in G_6$, there exists an embedding of τ in an L-slab. Similarly, for $L = 1$ and any graph $\tau \in G_5$, there exists an embedding of τ in a 1-slab. Finally, in Soteros (1992) [10] it was shown that for $L = 0$ and any planar graph $\tau \in G_4$, there exists an embedding of τ in the 0-slab, i.e., the square lattice Z^2. The arguments that led to Lemma 2.4 have been modified for an L-slab in [15] for the case $L > 0$ and in [10] for the case $L = 0$ and used to prove the following.

LEMMA 4.4. *For $L > 1$ and any fixed graph $\tau \in G_6$*

$$\lim_{n\to\infty} n^{-1} \log g_n(\tau, L) = \kappa(L); \tag{4.4}$$

for $L = 1$ and any fixed graph $\tau \in G_5$

$$\lim_{n\to\infty} n^{-1} \log g_n(\tau, 1) = \kappa(1); \tag{4.5}$$

and for $L = 0$ and any planar graph $\tau \in G_4$

$$\lim_{n\to\infty} n^{-1} \log g_n(\tau, 0) = \kappa(0). \tag{4.6}$$

In all three cases the limits are taken only through values of n for which $g_n(\tau, L) > 0$.

The result, corresponding to Lemma 2.5, that uniform embeddings in an L-slab have the same growth constant as self-avoiding walks in an L-slab,

$$\lim_{n\to\infty} (nf)^{-1} \log g_n(\tau, L; (n, n, ..., n)) = \kappa(L), \tag{4.7}$$

was proved for $L = 0$ in [10] and for $L > 0$ in Whittington and Soteros [15]. Similar arguments can also be used to show that, for $L \geq 1$,

$$\lim_{n\to\infty} N^{-1} \log g_N(\tau, L; \phi(n)) = \kappa(L), \tag{4.8}$$

where $g_N(\tau, L; \phi(n))$ is the number of embeddings of τ in an L-slab with edge distribution $\phi(n)$ and where the restrictions put on $\{\phi(n)\}_{n\geq 1}$ are similar to to those listed in Lemma 2.6.

Using arguments analogous to those which prove Lemma 3.2, the following pattern theorem holds for embeddings of graphs in an L-slab.

LEMMA 4.5. *Given any $m_1 > 0$ and $m_2 > 0$ and any pattern P which is an undirected version of a $K_{m_1,m_2}(L)$ pattern, let $g_n(\tau, L; \bar{P})$ be the number of n-edge embeddings of τ in an L-slab (up to translation) which do not contain the pattern P. For $L > 1$ and τ any graph in G_6,*

$$\limsup_{n\to\infty} n^{-1} \log g_n(\tau, L; \bar{P}) \leq \kappa(L; \bar{P}) < \kappa(L); \tag{4.9}$$

for $L = 1$ and τ any graph in G_5,

$$\limsup_{n\to\infty} n^{-1} \log g_n(\tau, L; \bar{P}) \leq \kappa(L; \bar{P}) < \kappa(1). \tag{4.10}$$

For each $L \geq 1$ it is possible to construct a tight trefoil pattern $T(L)$ so that if $T(L)^q$ appears in a cycle of an embedding, the embedding will surely be knotted. The tight trefoil pattern T introduced in Figure 2 is in fact suitable for all $L \geq 1$. Based on Lemma 4.3 with P an appropriate tight trefoil, the following result for unknotted polygons in an L-slab has been proved.

LEMMA 4.6 (TESI *et al* 1994 [4]). *Let* $p_n^o(L)$ *be the number (up to translation) of* n*-edge unknotted self-avoiding polygons in an* L*-slab. Then the limit in the following inequality exists and satisfies*

$$\lim_{n\to\infty} (2n)^{-1} \log p_{2n}^o(L) = \kappa_o(L) < \kappa(L) \tag{4.11}$$

for all $L \geq 1$, $\kappa_o(L)$ *is a concave function of* L,

$$\lim_{L\to\infty} \kappa_o(L) = \kappa_o, \tag{4.12}$$

and

$$\lim_{L\to\infty} \kappa(L) - \kappa_o(L) = \alpha \tag{4.13}$$

where α *is defined as in equation (3.9).*

The numerical evidence presented in [4,14] indicates that $\alpha(L) \equiv \kappa(L) - \kappa_o(L)$ is a decreasing function of L and hence the probability of a polygon being knotted increases as the height L of the slab decreases.

Using Lemma 4.6 and arguing as in the proof of Theorem 3.1, the following theorem for knots in graphs in an L-slab results.

THEOREM 4.1. *For* $L > 1$ $(L = 1)$, *suppose* $\tau \in G_6$ (G_5) *has no cut edges,*

$$\begin{aligned} \kappa_o(L) \leq \liminf_{n\to\infty} n^{-1} \log g_n^o(\tau, L) &\leq \limsup_{n\to\infty} n^{-1} \log g_n^o(\tau, L) \\ &\leq \kappa(\bar{T}^q, L) < \kappa(L) \end{aligned} \tag{4.14}$$

(where the limit is taken only through values of n *for which* $g_n^o(\tau, L) > 0$*) and hence the probability that an* n*-edge embedding of* τ *is knotted goes to unity at least as fast as*

$$1 - e^{-\beta(L)n + o(n)} \tag{4.15}$$

and no faster than

$$1 - e^{-\alpha(L)n + o(n)} \tag{4.16}$$

when $n \to \infty$, *with* $\beta(L) = \kappa(L) - \kappa(\bar{T}^q, L)$ *and* $\alpha(L) = \kappa(L) - \kappa_o(L)$.

If $\tau \in \tilde{G}_6$ $(\tilde{G}_5)$ *is eulerian*

$$\lim_{n\to\infty} (2n)^{-1} \log g_{2n}^o(\tau) = \kappa_o(L) \tag{4.17}$$

and hence the probability that an n*-edge embedding of* τ *is knotted goes to unity as*

$$1 - e^{-\alpha(L)n + o(n)} \tag{4.18}$$

when $n \to \infty$.

A result analogous to Theorem 3.2 also holds for embeddings of τ with a fixed edge distribution sequence.

5. Knots in graphs in an (L, M)-prism. An (L, M)-rectangular prism (or tube), is defined to be the sublattice of Z^3 induced by the vertex set $\{(x, y, z) \in Z^3 | 0 \leq y \leq L, 0 \leq z \leq M\}$. Let $c_n(L, M)$ be the number of self-avoiding walks confined to an (L, M)-prism, let $p_n(L, M)$ be the number (up to translation) of n-edge self-avoiding polygons confined to an (L, M)-prism, and given a graph $\tau \in G_6$, let $g_n(\tau, L, M)$ be the number (up to translation) of n-edge embeddings of τ in an (L, M)-prism.

The procedure followed in the last two sections to obtain results about knots in graphs no longer works for embeddings of graphs in an (L, M)-prism. In fact the procedure fails at the first step because there is no result, such as the result in Lemma 4.1 for walks and polygons in an L-slab, which equates the growth constants for self-avoiding walks and polygons in a prism. Instead as the following two results indicate the number of self-avoiding polygons in an (L, M)-prism is exponentially smaller than the number of self-avoiding walks in an (L, M)-prism.

LEMMA 5.1 (SOTEROS AND WHITTINGTON 1989 [16]). *The following limit exits*

$$\lim_{n \to \infty} n^{-1} \log c_n(L, M) \equiv \kappa(L, M). \tag{5.1}$$

LEMMA 5.2 (SOTEROS AND WHITTINGTON 1989 [16]). *The limit in the following inequality exists and satisfies*

$$\lim_{n \to \infty} (2n)^{-1} \log p_{2n}(L, M) \equiv \kappa_p(L, M) < \kappa(L, M). \tag{5.2}$$

A pattern theorem for self-avoiding walks in an (L, M)-prism was also proved in [16]. Let $C_b(L, M)$ be the set of self-avoiding walks completely contained in $D_b(L, M) = \{(x, y, z) \in Z^3 | 0 \leq x \leq b, 0 \leq y \leq L, 0 \leq z \leq M\}$ with one endpoint at the origin and the other at $(b, 0, 0)$. For any $b > 0$, a K_b pattern is defined to be any self-avoiding walk which can occur in some element of $C_b(L, M)$.

LEMMA 5.3 (SOTEROS AND WHITTINGTON 1989 [16]). *For any $b > 0$, let P be a K_b pattern, then*

$$\lim_{n \to \infty} n^{-1} \log c_n(L, M; \bar{P}) \equiv \kappa(L, M; \bar{P}) < \kappa(L, M). \tag{5.3}$$

With regards to embeddings of graphs in an (L, M)-prism, the following results indicates that if a graph has a cut edge then its growth constant is the same as that for self-avoiding walks, if a graph is eulerian then its growth constant is the same as that for self-avoiding polygons, otherwise it remains even to be proved that the growth constant exists.

LEMMA 5.4. *Given a graph $\tau \in G_6$ such that $g_n(\tau, L, M) > 0$ for some n, if τ has a cut edge*

$$\lim_{n \to \infty} n^{-1} \log g_n(\tau, L, M) = \kappa(L, M); \tag{5.4}$$

if τ is eulerian

$$\lim_{n\to\infty}(2n)^{-1}\log g_{2n}(\tau,L,M)=\kappa_p(L,M); \tag{5.5}$$

and if τ is not eulerian and has no cut edges

$$\begin{aligned}\kappa_p(L,M)\le\liminf_{n\to\infty}n^{-1}\log g_n(\tau,L,M)\le\quad&\limsup_{n\to\infty}n^{-1}\log g_n(\tau,L,M)\\ &<\kappa(L,M).\end{aligned} \tag{5.6}$$

Proof. In the case that τ has a cut edge an upper bound is obtained, just as in the proof of Lemma 2.4, by separating an embedding of τ into self-avoiding walks (each confined to an (L,M)-prism) to give

$$\begin{aligned}g_n(\tau,L,M)\quad&\le\sum_{\{m_i\mid\sum_{i=1}^{f}m_i=n\}}c_{m_1}(L,M)c_{m_2}(L,M)\cdots c_{m_f}(L,M)\\ &\le e^{\kappa(L,M)n+o(n)}.\end{aligned} \tag{5.7}$$

For the lower bound an argument similar to that used in [10, Theorem 3] is used. Fix a cut edge of τ. Obtain two rooted graphs, τ_L and τ_R, from τ by deleting the cut edge. Find an embedding of τ_L (τ_R) in an (L,M)-prism such that its root, the vertex which was connected to the designated cut edge of τ, is in the rightmost (leftmost) plane of the embedding. The two embeddings of τ_L and τ_R can now be concatenated together by concatenating the first vertex of an n-step ***unfolded*** walk (or bridge) to τ_L's root and then concatenating the walk's last vertex to τ_R's root (the concatenation of the walk to the roots is done using two short walks each with lengths less than $2(L+1)(M+1)$). This gives the lower bound

$$c_n^{\dagger}(L,M)\le g_{n+m}(\tau,L,M) \tag{5.8}$$

where m is the number of edges of the resulting embedding that were not initially part of the unfolded walk and $c_n^{\dagger}(L,M)$ is the number of n-step unfolded self-avoiding walks in an (L,M)-prism. It has been proved [17] (see also [7, section 8.2]) that

$$\lim_{n\to\infty}n^{-1}\log c_n^{\dagger}(L,M)=\kappa(L,M). \tag{5.9}$$

Hence taking logarithms, dividing by n, and letting $n\to\infty$ in equations (5.7) and (5.8) gives equation (5.4).

In the case that τ is eulerian, an argument similar to that used to obtain equation (3.13) is used. That is, an upper bound is obtained by separating τ into self-avoiding polygons and a lower bound is obtained by concatenating a polygon to an embedding of τ. Taking logarithms, dividing by n, and letting $n\to\infty$ in the resulting inequalities leads to equation (5.5).

For τ with no cut edges and not eulerian, for example τ a θ graph, a lower bound can be obtained by concatenating a polygon to an embedding of τ. This leads to the left-hand side of equation (5.6). Since every branch of τ is contained in a cycle, no branch of τ can contain either of the *filling patterns* depicted in [16, Figure 4] (these patterns cannot appear in any self-avoiding polygon). Thus equation (5.7) becomes

$$\begin{aligned} g_n(\tau, L, M) &\leq \sum{}' c_{m_1}(L, M; \bar{P_F}) c_{m_2}(L, M; \bar{P_F}) \cdots c_{m_f}(L, M; \bar{P_F}) \\ (5.10) \qquad &\leq e^{\kappa(L,M;\bar{P_f})n+o(n)} \end{aligned}$$

where P_F is the appropriate filling pattern and the primed sum is over the set $\{m_i | \sum_{i=1}^{f} m_i = n\}$. Equation (5.10) and Lemma 5.3 imply equation (5.6). □

To obtain results for embeddings of graphs with a fixed edge distribution sequence, a pattern theorem for self-avoiding polygons in an (L, M)-prism is needed. Due to Lemma 5.2, a pattern theorem for self-avoiding walks in a prism does not imply a pattern theorem for self-avoiding polygons. Instead, a separate pattern theorem for self-avoiding polygons in an (L, M)-prism is needed. The required pattern theorem is introduced here and a proof is given in the next section.

Define $\mathcal{P}(L, M)$ to be the set of self-avoiding polygons in an (L, M)-prism. Let $\mathcal{P}_b(L, M) \subseteq \mathcal{P}(L, M)$ be the set of self-avoiding polygons confined to the subgraph of the prism given by the vertex set $V_b = \{(x, y, z) \in Z^3 | 0 \leq x \leq b, 0 \leq y \leq L, 0 \leq z \leq M\}$. Given any $b > 0$, a $\tilde{K}_b$ pattern is defined to be a configuration (including occupied and unoccupied edges) of any element of $\mathcal{P}_b(L, M)$ in V_b with the edges in the $x = 0$ and $x = b$ planes excluded. Given a $\tilde{K}_b$ pattern P, define $p_n(L, M; \bar{P})$ to be the number (up to translation) of n-edge polygons confined to an (L, M)-prism in which the pattern P does not occur. The following lemma is a consequence of the pattern theorem that is proved in the next section.

LEMMA 5.5. *For any integer $b \geq 2$ and any $\tilde{K}_b$ pattern P,*

$$(5.11) \qquad \lim_{n \to \infty} n^{-1} \log p_n(L, M; \bar{P}) \equiv \kappa_p(L, M; \bar{P}) < \kappa_p(L, M).$$

The prism pattern theorems, Lemma 5.4 and Lemma 5.5, can be used to investigate the growth constant for embeddings of a graph with a fixed edge distribution sequence. Instead of exploring all possible types of graph embeddings, the focus is on several illustrative cases to show that the growth constant for graph embeddings in an (L, M)-prism can be highly dependent on the structure of the graph and the restrictions put on the edge distribution sequence. First some special graphs are defined. The *4-watermelon graph* is defined to be the graph with two vertices, v_1 and v_2, and four edges, each of the form $\{v_1, v_2\}$. A *dumbell graph* is the graph with two vertices of degree 3 and three edges, one edge from one vertex of the graph to the other and two loops, one from each vertex to itself. A *3-star* is a graph with one vertex of degree 3, 3 vertices of degree 1, and three edges, one from the vertex of degree 3 to each vertex of degree 1. The following theorem indicates that if the fast growing branches of τ form a chain with essentially free ends then the embeddings of τ have the same growth constant as self-avoiding walks; however, if instead there is a fast growing branch in a cycle or three fast growing branches form a 3-star the embeddings have less freedom or entropy than self-avoiding walks and the growth constant is reduced. Furthermore, if the fast growing branches of τ are contained in at most two loops of τ then the embeddings of τ have the same growth constant as self-avoiding polygons; however, if instead the fast growing branches are contained in at least three branches of a 4-watermelon graph the embeddings have less freedom or entropy and the growth constant is reduced.

LEMMA 5.6. *Given $L \geq 1$, $M \geq 1$, suppose that $\tau \in G_6$ is such that $g_n(\tau, L, M) > 0$ for some n and that $\{\phi(n)\}_{n\geq 1}$ is an edge distribution sequence for τ such that $g_N(\tau, L, M; \phi(n)) > 0$ for all $n \geq 1$ and for each i, $\phi_i(n)$ is either a constant or strictly increasing function of n.*

First suppose τ has a cut edge. Also suppose $\{\phi(n)\}_{n\geq 1}$ is such that all the fast growing branches of τ are cut edges and that when they are removed from τ, τ decomposes into components each of which was connected to at most two fast growing branches of τ. In this case

$$\lim_{n\to\infty} N^{-1} \log g_N(\tau, L, M; \phi(n)) = \kappa(L, M). \tag{5.12}$$

On the other hand, if $\{\phi(n)\}_{n\geq 1}$ is such that either there is a fast growing branch that is contained within a cycle or there are three or more fast growing branches connected to one branch point, then

$$\limsup_{n\to\infty} N^{-1} \log g_N(\tau, L, M; \phi(n)) < \kappa(L, M). \tag{5.13}$$

Now suppose τ has at least one cycle and $\{\phi(n)\}_{n\geq 1}$ is such that all fast growing branches are contained within an eulerian subgraph (not necessarily connected) of τ. Then in general

$$\limsup_{n\to\infty} N^{-1} \log g_N(\tau, L, M; \phi(n)) \leq \kappa_p(L, M). \tag{5.14}$$

Furthermore, if $\{\phi(n)\}_{n\geq 1}$ is such that all fast growing branches are contained in either of two loops of τ, then

$$\lim_{n\to\infty} N^{-1} \log g_N(\tau, L, M; \phi(n)) = \kappa_p(L, M). \tag{5.15}$$

However, if τ has a subgraph which is homeomorphic to the 4-watermelon graph and if $\{\phi(n)\}_{n\geq 1}$ is such that at least three of the watermelon branches each contain a fast growing branch of τ, then

$$\limsup_{n\to\infty} N^{-1} \log g_N(\tau, L, M; \phi(n)) < \kappa_p(L, M). \tag{5.16}$$

Proof. Let τ be any graph in G_6 such that $g_n(\tau, L, M) > 0$ for some n and that $\{\phi(n)\}_{n\geq 1}$ is an edge distribution sequence for τ such that $g_N(\tau, L, M; \phi(n)) > 0$ for all $n \geq 1$.

First suppose τ has a cut edge. Since $g_N(\tau, L, M; \phi(n)) \leq g_N(\tau, L, M)$, equation (5.4) in Lemma 5.4 gives the upper bound

$$g_N(\tau, L, M; \phi(n)) \leq e^{\kappa(L,M)N+o(N)}. \tag{5.17}$$

Next suppose $\{\phi(n)\}_{n\geq 1}$ is such that all the fast growing branches of τ are cut edges and that when they are removed from τ, τ decomposes into components each of which was connected to at most two fast growing branches of τ. In this case a lower bound can be obtained for $g_N(\tau, L, M; \phi(n))$ in terms of $\kappa(L, M)$ as follows. Suppose τ has f' growing branches. Remove each fast growing branch from τ, this results in a disconnected graph with $f' + 1$ components, $\tau_1, ..., \tau_{f'+1}$, where by relabelling as appropriate we can assume that the ith fast growing branch of τ connects component τ_i to component τ_{i+1} in τ. Then arguing as in the proof of Lemma 5.4, by concatenating together elements of an alternating sequence of component embeddings and unfolded walks the following lower bound is obtained

$$c^{\dagger}_{\phi_1(n)-m_1} \cdots c^{\dagger}_{\phi_{f'}(n)-m_{f'}} \leq g_N(\tau, L, M; \phi(n)) \tag{5.18}$$

where $m_1, ..., m_{f'}$ are independent of n. Since $\lim_{n\to\infty} \frac{\sum_{i=1}^{f'} \phi_i(n)}{N} = 1$, this lower bound and equation (5.17) leads to equation (5.12).

Now suppose $\{\phi(n)\}_{n\geq 1}$ is such that either there is a fast growing branch of τ that is contained within a cycle or there are three or more fast growing branches connected to one branch point. As noted in the proof of Lemma 5.4, any branch of τ which is contained in a cycle cannot contain the filling pattern P_F. Thus if the ith branch of τ is a fast growing branch that is contained within a cycle

$$\begin{aligned} g_N(\tau, L, M; \phi(n)) &\leq c_{\phi_i(n)}(L, M; \bar{P_F}) \prod_{j=1, j\neq i}^{f} c_{\phi_j(n)}(L, M) \\ &\leq e^{(N-\phi_i(n))\kappa(L,M)+\phi_i(n)\kappa(L,M;\bar{P_F})+o(N)}. \end{aligned} \tag{5.19}$$

Taking logarithms, dividing by N, and letting $n \to \infty$ in this equation gives equation (5.13). Similarly, if three or more branches of τ are joined at a single branch point then at least one of these branches cannot contain the filling pattern P_F (if all three contained P_F then at least two of the branches will intersect each other at a vertex other than the common branch point). In this case, let i, $i + 1$ and $i + 2$ be the labels of the three fast growing branches

$$g_N(\tau, L, M; \phi(n)) \leq \left[\sum_{j=0}^{2} c_{\phi_{i+j}(n)}(L, M; \bar{P_F})\right] \prod_{m=1, m\neq i+j} c_{\phi_m(n)}(L, M)$$

$$\leq e^{(N-\phi_k(n))\kappa(L,M)+\phi_k(n)\kappa(L,M;\bar{P_F})+o(N)} \tag{5.20}$$

where $k \in \{i, i+1, i+2\}$ and $\lim_{n\to\infty} \frac{\phi_k(n)}{N} = \min_{0\leq j\leq 2}\{\lim_{n\to\infty} \frac{\phi_{i+j}(n)}{N}\}$. Taking logarithms, dividing by N, and letting $n \to \infty$ in this equation again gives equation (5.13).

If the only fast growing branches of τ are in an eulerian subgraph (not necessarily connected) η of τ, then by separating the eulerian subraph into f_p self-avoiding polygons and the remaining branches of τ into f' self-avoiding walks the following upper bound is obtained,

$$\begin{aligned} g_N(\tau, L, M; \phi(n)) &\leq R_\eta(N) \left[\prod_{i=1}^{f'} c_{\phi_i(n)}(L, M)\right] \prod_{j=1}^{f'} p_{m_j}(L, M) \\ &\leq e^{\kappa_p(L,M)N+o(N)} \qquad (5.21) \end{aligned}$$

where $R_\eta(N)$ is as defined in equation (3.19), m_i is the number of edges in η's ith self-avoiding polygon, and the fact that $\lim_{n\to\infty} \frac{\phi_i(n)}{N} = 0$ for $i = 1, ..., f'$ has been used to obtain the rightmost inequality. This leads to equation (5.14).

If τ is a figure eight graph, then a lower bound can be obtained for $g_n(\tau, L, M; \phi(n))$ in terms of $\kappa_p(L, M)$ by taking any two polygons with appropriate numbers of edges and then concatenating them (using $O(LM)$ additonal edges) together to create an embedding of a figure eight graph. The resulting lower bound combined with equation (5.14) leads to equation (5.15). Furthermore, if, for example, $\{\phi(n)\}_{n\geq 1}$ is such that τ has at most two fast growing branches each of which forms a loop of τ (an example is the case that τ is a dumbbell graph in which its cut edge is not fast growing), then the argument just described for a figure eight graph works in a similar way to give equation (5.15).

If τ is a 4-watermelon graph, an upper bound for $g_n(\tau, L, M; \phi(n))$ can be obtained by separating τ into two doubly rooted self-avoiding polygons. There are three ways to pair off the branches of τ to form polygons. Given an embedding of τ there is always at least one way to pair off the branches so that one of the two resulting polygons does not contain the filling pattern $P_{\hat{F}}$ depicted in Figure 3 (the pattern is defined so that the branch points of the watermelon cannot be within the pattern and so that it cannot appear in more than two branches of the 4-watermelon graph). Given an embedding of τ, relabel the four branches so that branch 1 and 2 are used to form a polygon which does not contain pattern $P_{\hat{F}}$ and let this be the last in the sequence of polygons obtained from τ. Thus,

$$\begin{aligned} g_N(\tau, L, M; \phi(n)) &\leq N^4 p_{\phi_3(n)+\phi_4(n)}(L, M) p_{\phi_1(n)+\phi_2(n)}(L, M; \bar{P_{\hat{F}}}) \\ &\leq e^{(N-\phi_1(n)-\phi_2(n))\kappa_p(L,M)+(\phi_1(n)+\phi_2(n))\kappa_p(L,M;\bar{P_{\hat{F}}})+o(N)}. \qquad (5.22) \end{aligned}$$

If at least three of the branches are fast growing then at least one of the branches 1, 2 will be fast growing. Taking logarithms, dividing by N, letting

$n \to \infty$ in the above equation gives equation (5.16) for the case that τ is a 4-watermelon graph. This same argument works if τ has a subgraph which is homeomorphic to a 4-watermelon graph and $\{\phi(n)\}_{n\geq 1}$ is such that each of three of the branches of the watermelon contain fast growing branches of τ. There are also other types of graphs and edge distribution sequences for which equation (5.16) will hold. □

Note that equation (5.13) was shown for $\phi(n) = (n, n, ..., n)$ (i.e. uniform embeddings) in [15].

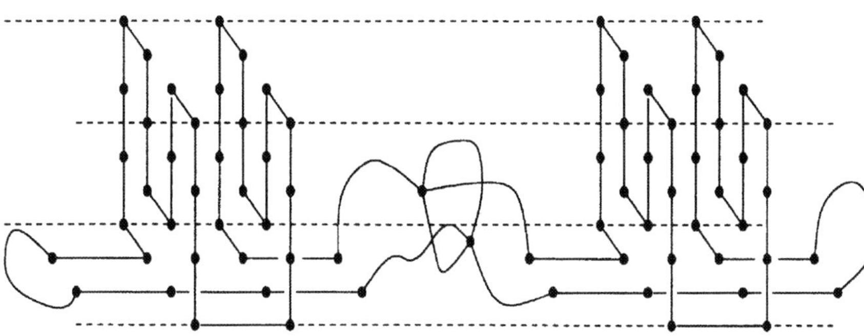

FIG. 3. *The bold pattern (either on the left or the right of the prism) is a filling pattern $\hat{P}_F$ which cannot appear in more than two branches of a 4-watermelon graph in a (3,3)-prism. A similar filling pattern exists for an (L, M)-prism.*

Next the probability that a graph embedding is knotted is investigated. Define $p_n^o(L, M)$ to be the number (up to translation) of unknotted self-avoiding polygons in an (L, M)-prism. Similarly, define $g_n^o(\tau, L, M)$ to be the number (up to translation) of unknotted embeddings of τ in an (L, M)-prism. A tight trefoil $\tilde{K}_b$ pattern is now needed. Consider T in Figure 2, suppose the left most vertex in T is the origin. T can be made into a $\tilde{K}_5$ pattern by adding in 5 connected edges starting at $(0, 0, -1)$ and ending at $(5, 0, -1)$ (or starting at $(0, -2, 0)$ and ending at $(0, -2, 5)$) and such a pattern can fit in an (L, M)-prism for $L \geq 3$, $M \geq 2$ (or $L \geq 4$,$M \geq 1$). Using the pattern theorem for self-avoiding polygons in a prism (Lemma 5.5) and a concatenation argument, the following result can be proved for any L and M for which a tight trefoil pattern in an (L, M)-prism exists.

LEMMA 5.7. *For L and M such that the (L, M)-prism contains a tight trefoil $\tilde{K}_b$ pattern for some $b > 0$,*

$$\lim_{n\to\infty} n^{-1} \log p_n^o(L, M) \equiv \kappa_o(L, M) < \kappa_p(L, M) \tag{5.23}$$

and hence the probability that a self-avoiding polygon in an (L, M)-prism is knotted goes to unity as

$$1 - e^{-\alpha(L,M)n+o(n)} \tag{5.24}$$

when $n \to \infty$, *with* $\alpha(L,M) = \kappa_p(L,M) - \kappa_o(L,M)$.

It is now possible to obtain results about knots in graphs in an (L,M)-prism using Lemma 5.7 and focussing on the cases in which the embeddings of a graph have growth constant equal to $\kappa_p(L,M)$.

THEOREM 5.1. *Let* L *and* M *be as in Lemma 5.7 and* T *be an appropriate tight trefoil* $\tilde{K}_b$ *pattern for some* $b > 0$. *Given eulerian* $\tau \in G_6$ *such that* $g_n(\tau, L, M) > 0$ *for some* $n > 0$,

$$\begin{aligned} \kappa_o(L,M) &\leq \liminf_{n\to\infty} n^{-1}\log g_n^o(\tau,L,M) \\ &\leq \limsup_{n\to\infty} n^{-1}\log g_n^o(\tau,L,M) \leq \kappa_p(L,M;\bar{T}^q) < \kappa_p(L,M). \end{aligned} \tag{5.25}$$

Further, given any $\tau \in G_6$ *and an edge distribution sequence such that equation (5.15) is true and there exists an unknotted embedding of* τ *with edge distribution* $\phi(1)$,

$$\begin{aligned} \kappa_o(L,M) &\leq \liminf_{n\to\infty} N^{-1}\log g_N^o(\tau,L,M;\phi(n)) \\ &\leq \limsup_{n\to\infty} N^{-1}\log g_N^o(\tau,L,M;\phi(n)) \\ &\leq \kappa_p(L,M;\bar{T}^q) < \kappa_p(L,M). \end{aligned} \tag{5.26}$$

In both these cases the probability that an appropriate n*-edge embedding of* τ *in an* (L,M)*-prism is knotted goes to unity at least as fast as*

$$1 - e^{-\beta(L,M)N + o(N)} \tag{5.27}$$

and no faster than

$$1 - e^{-\alpha(L,M)N + o(N)} \tag{5.28}$$

when $n \to \infty$, *with* $\beta(L,M) \equiv \kappa_p(L,M) - \kappa_p(L,M;\bar{T}^q)$ *and* $\alpha(L,M)$ *as in Lemma 5.7.*

For $\tau \in \tilde{G}_6$, *these results can be strengthened to:*

$$\lim_{n\to\infty} n^{-1}\log g_n^o(\tau,L,M) = \kappa_o(L,M) \tag{5.29}$$

for eulerian graphs and

$$\lim_{n\to\infty} N^{-1}\log g_N^o(\tau,L,M;\phi(n)) = \kappa_o(L,M) \tag{5.30}$$

for any τ *and edge distribution sequence such that equation (5.15) is true. In both these cases the probability that an appropriate* n*-edge embedding of* τ *is knotted goes to unity as*

$$1 - e^{-\alpha(L,M)N + o(N)} \tag{5.31}$$

when $n \to \infty$.

Proof. Obtain bounds for $g_n^o(\tau,L,M)$ and $g_N^o(\tau,L,M;\phi(n))$ as in the proofs of equations (5.5) and (5.15), except instead of obtaining bounds in terms of arbitrary self-avoiding polygons either self-avoiding polygons which do not contain T^q or unknotted self-avoiding polygons (depending on whether $\tau \in G_6 - \tilde{G}_6$ or $\tau \in \tilde{G}_6$) are used. □

6. A pattern theorem for self-avoiding polygons in a (L, M)-prism. In this section a transfer matrix approach, the same as that used by Alm and Janson in their study of self-avoiding walks in one-dimensional lattices [18], is used to prove a pattern theorem for self-avoiding polygons in a rectangular prism.

An (L, M)-prism is equivalent to $Z \times H(L, M)$, where $H(L, M)$ is the finite subgraph of Z^2 induced by the vertex set $\{(y, z) \in Z^2 | 0 \leq y \leq L, 0 \leq z \leq M\}$, and hence, in the terminology of [18], an (L, M)-prism is a one-dimensional lattice. An (L, M)-prism can also be thought of as an alternating sequence of *hinges* and *sections* where for any $i \in Z$ the ith hinge, $H_i(L, M)$, is defined to be the subgraph of the prism induced by the vertex set $\{(i, y, z) \in Z^3 | 0 \leq y \leq L, 0 \leq z \leq M\}$ and the ith section, $S_i(L, M)$, is defined to be the set of edges which join $H_{i-1}(L, M)$ to $H_i(L, M)$ in the prism.

A self-avoiding polygon in an (L, M)-prism is said to span m sections of the prism if the edges of the self-avoiding polygon are contained in $H_{i-1}(L, M) \cup S_i(L, M) \cup H_i(L, M) \cup ... \cup S_{i+m-1}(L, M) \cup H_{i+m-1}(L, M)$ for some $i \in Z$. Clearly, given an n-edge self-avoiding polygon in an (L, M)-prism there exists some $m \in \{1, ..., \frac{n-2}{2}\}$ such that the polygon spans m sections of the prism; define m to be the *span* of the given self-avoiding polygon.

Without loss of generality, assume that the first hinge and section of any self-avoiding polygon in an (L, M)-prism are respectively $H_0(L, M)$ and $S_1(L, M)$ and define the *last* section and hinge to be $S_m(L, M)$ and $H_m(L, M)$, respectively, where m is the span of the polygon. Furthermore, for any polygon in an (L, M)-prism the edges of the polygon can be directed and ordered in the following way: consider the vertices joined by an edge of the polygon to the polygon's bottom vertex (the first vertex of the polygon in a lexicographic ordering of the coordinates of the polygon's vertices), of these choose the smallest (lexicographically) vertex and direct an edge from the bottom vertex to this vertex. This edge is considered the first in the ordering of the edges. The ordering and orientation of the remaining edges continues in a cyclic fashion around the polygon. Given a self-avoiding polygon in an (L, M)-prism with span m, define a *pattern* with span k of the polygon to be the polygon's *configuration* in a sublattice of the form $S_i(L, M) \cup H_i(L, M) \cup ... \cup H_{i+k-2}(L, M) \cup S_{i+k-1}(L, M)$ for $1 \leq i \leq m - k + 1$. The polygon's configuration in such a sublattice of the prism consists of the sublattice, the set of directed edges of the polygon in the sublattice, and if there are e edges of the polygon in the sublattice then they are ordered from 1 to e according to their ordering in the polygon. We say the pattern of span k corresponding to a polygon's configuration in $S_i(L, M) \cup H_i(L, M) \cup ... \cup H_{i+k-2}(L, M) \cup S_{i+k-1}(L, M)$ *occurs* at the ith section of the polygon. Figure 4 depicts a self-avoiding polygon in a $(0, 4)$-prism and its associated patterns with span 2. Given $k \geq 2$, define $\Pi(k)$ to be the set of distinct patterns with span k that occur in at least

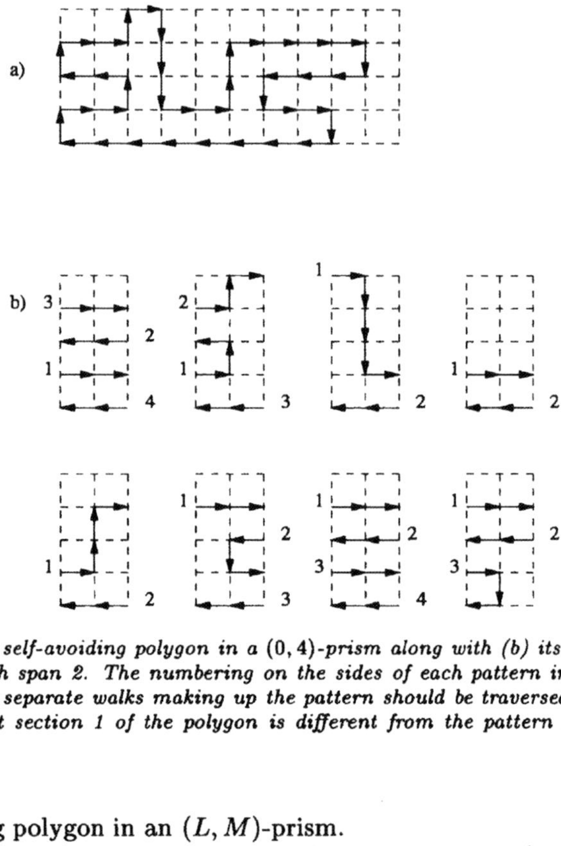

FIG. 4. *(a) A self-avoiding polygon in a* $(0,4)$*-prism along with (b) its eight associated patterns with span 2. The numbering on the sides of each pattern indicates the order in which the separate walks making up the pattern should be traversed. Thus the pattern occuring at section 1 of the polygon is different from the pattern occuring at section 7.*

one self-avoiding polygon in an (L,M)-prism.

Consider $\hat{\mathcal{P}}(L,M)$, the set of self-avoiding polygons in an (L,M)-prism with the following properties: the only edge in $H_0(L,M)$ is the edge between $(0,0,0)$ and $(0,0,1)$; the only edges in $S_1(L,M)$ are the edge between $(1,0,0)$ and $(0,0,0)$ and the edge between $(0,0,1)$ and $(1,0,1)$; the only edges in $S_m(L,M)$ are the edge between $(m,0,0)$ and $(m-1,0,0)$ and the edge between $(m-1,0,1)$ and $(m,0,1)$, for m the span of the polygon; the only edge in $H_m(L,M)$ is the edge between $(m,0,0)$ and $(m,0,1)$; and if the edges of the polygon are oriented in a cyclic fashion starting from $(0,0,0)$ to $(0,0,1)$, then the orientation of the edge in $H_m(L,M)$ is from $(m,0,1)$ to $(m,0,0)$. Let $\hat{p}_n(L,M)$ be the number of such self-avoiding polygons with n-edges.

The following result indicates that the growth constant for $\hat{p}_n(L,M)$ is the same as the growth constant for $p_n(L,M)$.

LEMMA 6.1.

$$\lim_{n\to\infty} n^{-1}\log \hat{p}_n(L,M) = \kappa_p(L,M). \tag{6.1}$$

Also, for any integer $k \geq 1$ *and any pattern* $\pi \in \Pi(k)$*,* π *can occur in an* *element of* $\hat{\mathcal{P}}(L, M)$.

Proof. Clearly

$$\hat{p}_n(L, M) \leq p_n(L, M). \tag{6.2}$$

To get a lower bound for $\hat{p}_n(L, M)$, start with any n-edge polygon in an (L, M)-prism. Orient and order the edges of the polygon in a cyclic fashion as described previously. It is always possible to concatenate an oriented polygon (with on the order of $(2L + 2M)$ edges) to the first edge and another oriented polygon (with on the order of $(2L + 2M)$ edges) to the last edge in the rightmost plane, so that the resulting polygon is an element of $\hat{\mathcal{P}}(L, M)$. Thus

$$p_n(L, M) \leq \hat{p}_{n+O(2L+2M)}(L, M) \tag{6.3}$$

where the function $O(2L + 2M)$ does not depend on n. Taking logarithms, dividing by n, and then letting n go to infinity in equations (6.3) and (6.2) results in equation (6.1).

Also, given a pattern $\pi \in \Pi(k)$ there exists a self-avoiding polygon in an (L, M)-prism such that π occurs in the self-avoiding polygon. Fix such a self-avoiding polygon and now, as in the argument leading to the lower bound for $\hat{p}_n(L, M)$, concatenate polygons to the first and last hinge of the initial polygon to create an element of $\hat{\mathcal{P}}(L, M)$. Thus an element of $\hat{\mathcal{P}}(L, M)$ has been created in which the pattern π occurs. □

For any integer $k \geq 1$ and given a pattern $\pi \in \Pi(k)$, define $p_n(N, M; \bar{\pi})$ to be the number (up to translation) of n-edge polygons in an (L, M)-prism that do not have π as a pattern. Similarly, define $\hat{p}_n(N, M; \bar{\pi})$ to be the number of n-edge polygons of $\hat{\mathcal{P}}(L, M)$ that do not have π as a pattern. The following is a pattern theorem for self-avoiding polygons in an (L, M)-prism.

THEOREM 6.1. *For any integer $k \geq 2$ and any pattern $\pi \in \Pi(k)$,*

$$\begin{aligned} \lim_{n\to\infty} n^{-1} \log p_n(L, M; \bar{\pi}) = \lim_{n\to\infty} n^{-1} \log \hat{p}_n(L, M; \bar{\pi}) \quad &\equiv \quad \kappa_p(L, M; \bar{\pi}) \\ &< \quad \kappa_p(L, M). \end{aligned} \tag{6.4}$$

Proof. Note first that if the second limit in the above equation exists then the arguments of Lemma 6.1 imply that

$$\lim_{n\to\infty} n^{-1} \log p_n(L, M; \bar{\pi}) = \lim_{n\to\infty} n^{-1} \log \hat{p}_n(L, M; \bar{\pi}). \tag{6.5}$$

In particular, given the pattern π, one can always convert a polygon in an (L, M)-prism into an element of $\hat{P}(L, M)$ by a concatenation process which ensures that if π does not occur in the original polygon then it does not occur in the final polygon. Thus in the remainder of the proof the focus will be on the set of polygons in $\hat{P}(L, M)$.

Given $k \geq 2$, consider the set of all possible patterns with span k, $\Pi(k)$. Each such pattern can be considered as a polygon configuration of

$S_1(L,M)\cup H_1(L,M)\cup\ldots\cup H_{k-1}(L,M)\cup S_k(L,M)$. Since these patterns are contained in a finite subgraph of the lattice there is clearly a finite number of such patterns. Order them from 1 to $|\Pi(k)|$. For the ith pattern, define e_i to be the number of polygon edges contained in $S_1(L,M)\cup H_1(L,M)$. We can construct a transfer matrix $G(x)=(g_{i,j}(x))$ in the following way:

$$g_{i,j}(x)=\begin{cases} x^{e_i} & \begin{array}{c}\text{if the configuration of pattern } i \text{ on}\\ S_2(L,M)\cup H_2(L,M)\cup\ldots\cup H_{k-1}(L,M)\cup S_k(L,M)\\ \text{equals the configuration of pattern } j \text{ on}\\ S_1(L,M)\cup H_1(L,M)\cup\ldots\cup H_{k-2}(L,M)\cup S_{k-1}(L,M)\end{array}\\ 0 & \text{otherwise.}\end{cases}$$

(6.6)

Define the pattern $\phi\in\Pi(k)$ to be the pattern with the following configuration: the first polygon edge goes from (0,0,1) to (1,0,1), the second edge goes from (1,0,1) to (2,0,1),..., the kth edge goes from ($k-1$,0,1) to (k,0,1), the $k+1$st edge goes from (k,0,0) to ($k-1$,0,0), the $k+2$nd edge goes from ($k-1$,0,0) to ($k-2$,0,0), ..., the ($2k$)th edge goes from (1,0,0) to (0,0,0). Let ϕ also represent the index of ϕ in the labelling of all patterns in $\Pi(k)$.

Consider a sequence of $r\geq 2$ patterns in $\Pi(k)$ of the form ϕ, π_{i_2}, π_{i_3}, ..., $\pi_{i_{r-1}}$, ϕ such that $g_{\phi,i_2}(x)\neq 0$, $g_{i_{r-1},\phi}(x)\neq 0$, and $g_{i_j,i_{j+1}}(x)\neq 0$ for $2\leq j\leq r-2$. Then this sequence defines a pattern with span $r+k-1$ in which pattern ϕ occurs at the 1st and rth section and pattern π_{i_j} occurs at the jth section, for $j=2,\ldots,r-1$. The number of edges in this pattern is $2e_\phi+\sum_{j=2}^{r-1}e_{i_j}+2(k-1)$ where $e_\phi=2$ and the weight associated with this pattern in $(G(x)^{r-1})_{\phi,\phi}$ is $x^{e_\phi+\sum_{j=2}^{r-1}e_{i_j}}$. (The notation $(A)_{i,j}$ refers to the i,jth element of matrix A.)

Any pattern with span $r+k-1$ in which ϕ occurs at the 1st and rth section can be decomposed into a sequence of r patterns from $\Pi(k)$ as above. Note that for $r=2,\ldots,k$, the only pattern with span $r+k-1$ in which ϕ occurs at the 1st and rth section is the pattern which has ϕ occuring at each section. Also note that given $r\geq k$, a pattern with span $r+k-1$ and $2e_\phi+\sum_{j=2}^{r-1}e_{i_j}+2(k-1)$ edges corresponds to a self-avoiding polygon in $\hat{\mathcal{P}}(L,M)$ with span $r-k+1$ and $2e_\phi+\sum_{j=2}^{r-1}e_{i_j}-2(k-2)$ edges. Such a polygon is constructed from the pattern by deleting the $2(k-1)$ polygon edges of ϕ in $S_1(L,M)\cup H_1(L,M)\cup\ldots\cup S_{k-1}(L,M)\cup H_{k-1}(L,M)$, deleting the $2(k-1)$ polygon edges of ϕ in $H_r(L,M)\cup S_{r+1}(L,M)\cup H_{r+1}(L,M)\cup\ldots\cup S_{r+k-1}(L,M)$, and then adding edges from $(k-1,0,0)$ to $(k-1,0,1)$ and from $(r,0,1)$ to $(r,0,0)$. Translating the polygon so that the vertex $(k,0,0)$ is translated to $(0,0,0)$ results in a self-avoiding polygon in $\hat{\mathcal{P}}(L,M)$ with the properties as listed above. Figure 5 shows an example.

Thus the generating function $F(x)=\sum_{n=2}\hat{p}_{2n}(L,M)x^{2n}$ satisfies the

FIG. 5. *A pattern (on the left) starting and ending with $\phi \in \Pi(4)$ in a $(0,4)$-prism along with its associated self-avoiding polygon (on the right) in $\hat{\mathcal{P}}(0,4)$.*

following:

$$x^{2(k-3)}F(x) = (G(x)^{k-1})_{\phi,\phi} + (G(x)^k)_{\phi,\phi} + \ldots = (G(x)^{k-1}(I-G(x))^{-1})_{\phi,\phi} \tag{6.7}$$

where each non-zero term on the right hand side has a factor of x^{2k-2} (x^{2k-2} is the smallest power of x associated with a sequence of k correctly connected patterns with span k). Note that

$$(G(x)^{k-1}(I-G(x))^{-1})_{\phi,\phi} = \frac{\det(G(x)^{k-1})\det((I-G(x))(G(x)^{k-1})^{-1};\phi,\phi)}{\det(I-G(x))}. \tag{6.8}$$

Given any $x > 0$, it can be shown that for any pair of patterns i,j there exists an integer m such that $(G(x)^m)_{i,j} > 0$. (To see this, start with a polygon in $\hat{\mathcal{P}}(L,M)$ in which pattern i occurs and concatenate it to a polygon in $\hat{\mathcal{P}}(L,M)$ in which pattern j occurs. This yields a polygon in $\hat{\mathcal{P}}(L,M)$ in which pattern i occurs at some section and pattern j occurs at a later section. From this obtain a sequence of m correctly connected patterns starting with pattern i and ending in pattern j.) Also, $(G(x))_{\phi,\phi} = x^2 > 0$ for $x > 0$. Thus for $x > 0$, $G(x)$ is an irreducible and aperiodic matrix and Frobenius theory [19] implies that the spectral radius, $\rho(x)$, of $G(x)$ is a simple root of $\det(\lambda I - G(x))$, that $G(x)$ has a strictly positive eigenvector associated with $\rho(x)$, and that $\rho(x)$ is the only eigenvalue of modulus $\rho(x)$. $\rho(0) = 0$ and $\rho(x)$ is an unbounded, increasing, continuous function on $[0,\infty)$, hence there exists a unique $x_o > 0$ such that $\rho(x_o) = 1$. From equation (6.8), $F(x)$ has poles only when $\frac{1}{\det(I-G(x))}$ has poles, that is, when 1 is an eigenvalue of $G(x)$. Thus based on the results and arguments of [18, Lemma 9 and Theorem 3] $F(x)$ is analytic for $|x| < x_o$ and has one simple pole when $|x| = x_o$, namely, $x = x_o$. In particular, as $x \to x_o$

$$\begin{aligned}(x_o - x)F(x) &= (x_o - x)x^{6-2k}(G(x)^{k-1}(I-G(x))^{-1})_{\phi,\phi} \\ &\to x_o^{6-2k}(\zeta^T G'(x_o)\eta)^{-1}(\eta)_\phi(\zeta^T)_\phi > 0\end{aligned} \tag{6.9}$$

where η and ζ^T are strictly positive eigenvectors of $G(x_o)$ associated with $\rho(x_o) = 1$ and normalized so that $\zeta^T\eta = 1$. Thus, there exists $\alpha > 0$ such that

$$\hat{p}_n(L,M) = \alpha x_o^{-n} + o(x_o^{-n}) \qquad \text{as } n \to \infty \tag{6.10}$$

and taking $\kappa_p(L, M) = -\log x_o$ we obtain

$$\lim_{n\to\infty} n^{-1} \log \hat{p}_n(L, M) = \kappa_p(L, M). \tag{6.11}$$

Let $\pi \in \Pi(k)$ be a specific pattern, $\pi \neq \phi$. Let π represent the label of pattern π in $\Pi(k)$. Consider the generating function $\bar{F}(x) = \sum_{n=2} \hat{p}_{2n}(N, M; \bar{\pi}) x^{2n}$. Then clearly

$$x^{2(k-3)}\bar{F}(x) = (\bar{G}(x)^{k-1})_{\phi,\phi} + (\bar{G}(x)^k)_{\phi,\phi} + \ldots = (\bar{G}(x)^{k-1}(I - \bar{G}(x))^{-1})_{\phi,\phi} \tag{6.12}$$

where $\bar{G}(x)$ is obtained from $G(x)$ by deleting its πth row and column. The argument goes through just as before so that there exists $\bar{x}_o > 0$ and $\bar{\alpha} > 0$ such that

$$\hat{p}_n(N, M; \bar{\pi}) = \bar{\alpha}\bar{x}_o^{-n} + o(\bar{x}_o^{-n}) \qquad \text{as } n \to \infty \tag{6.13}$$

and the spectral radius, $\bar{\rho}(\bar{x}_o)$, of $\bar{G}(\bar{x}_o)$ equals 1. Now consider the matrix $G_\pi(x)$ obtained from $G(x)$ by replacing the πth row and column by a row and column of zeroes. Then the spectral radius, $\rho_\pi(x)$, of $G_\pi(x)$ equals $\bar{\rho}(x)$. Furthermore, $G_\pi(x) \leq G(x)$ and at least one element of $G_\pi(x)$ is strictly less than the corresponding element of $G(x)$. Frobenius theory [19] then implies that $\rho_\pi(x) < \rho(x)$ and hence for $x = x_o$, $\rho_\pi(x_o) < \rho(x_o) = 1$ and therefore $\bar{x}_o > x_o$. Thus,

$$\lim_{n\to\infty} n^{-1} \log \hat{p}_n(N, M; \bar{\pi}) = -\log \bar{x}_o < \kappa_p(L, M). \tag{6.14}$$

If $\pi = \phi$, redefine ϕ so that the y-coordinate of each polygon vertex is N instead of 0 and then a similar argument to that given above will yield equation (6.14). □

We note that it is possible to generalize this result to deal with other one dimensional lattices or to deal with patterns on the scale of hinges or sections (see [18] for such generalizations in the case of self-avoiding walks). In a subsequent work [20], questions related to the average number of times a particular pattern occurs in a self-avoiding polygon in a prism will be addressed.

Acknowledgements. The author is grateful to Neal Madras, Carla Tesi, and Stuart Whittington for helpful conversations, and to NSERC of Canada for financial support. Thanks also go to the I.M.A. for their hospitality and support while I was in residence there and to the organizers of the June 1997 I.M.A. workshop; during the period of time I was at the I.M.A. significant progress was made on this work.

REFERENCES

[1] D.W. Sumners and S.G. Whittington, *Knots in self-avoiding walks*, J. Physics A: Math. Gen., 21 (1988), pp. 1689–1694.

[2] N. Pippenger, *Knots in random walks*, Discrete Applied Math., 25 (1989), pp. 273–278.

[3] C.E. Soteros, D. W. Sumners and S. G. Whittington, *Entanglement complexity of graphs in Z^3*, Proc. Camb. Phil. Soc., 111 (1992), pp. 75–91.

[4] M.C. Tesi, E.J. Janse van Rensburg, E. Orlandini and S.G. Whittington, *Knot probability for lattice polygons in confined geometries*, J. Phys A: Math. Gen., 27 (1994), pp. 347–360.

[5] C.E. Soteros, *Random knots in uniform branched polymers*, Mathematical Modelling and Scientific Computing, 2, Section B (1993), pp. 747–752. (Proceedings of the Eighth International Conference on Mathematical and Computer Modelling. April 1-4, 1991, University of Maryland, College Park, Maryland)

[6] J.M. Hammersley and K.W. Morton, *Poor man's Monte Carlo*, J. R. Stat. Soc. B, 16 (1954), pp. 23–38.

[7] N. Madras and G. Slade, *The Self-Avoiding Walk*, (1993), (Boston: Birkhäuser)

[8] J.M. Hammersley, *The number of polygons on a lattice*, Proc. Camb. Phil. Soc., 57 (1961), pp. 516–523.

[9] J.M. Hammersley and S.G. Whittington, *Self-avoiding walks in wedges*, J. Phys. A: Math. Gen., 18 (1985), pp. 101–111.

[10] C.E. Soteros, *Adsorption of uniform lattice animals with specified topology*, J. Phys. A: Math. Gen., 25 (1992), pp. 3153–3173.

[11] C.E. Soteros, D.W. Sumners and S.G. Whittington, *Linked p-spheres in Z^d*, (1997), *in preparation*

[12] J.H. Conway and C.McA. Gordon, *Knots and links in spatial graphs*, J. Graph Theory, 7 (1983), pp. 445–453.

[13] H. Kesten, *On the number of self-avoiding walks*, J. Math. Phys., 4 (1963), pp. 960–969.

[14] M.C. Tesi, E.J. Janse van Rensburg, E. Orlandini and S.G. Whittington, *Topological entanglement complexity of polymer chains in confined goemetries*, (1997), IMA Vol. in Math. and its Appl. (Springer-Verlag) this volume.

[15] S.G. Whittington and C.E. Soteros, *Uniform branched polymers in confined geometries*, Macromolecular Reports, A29(Suppl. 2) (1992), pp. 195–199.

[16] C.E. Soteros and S.G. Whittington, *Lattice models of branched polymers: effects of geometrical constraints*, J. Phys A: Math. Gen., 22 (1989), pp. 5259–5270.

[17] S.G. Whittington, *Statistical Mechanics of Polymer Solutions and Polymer Adsorption*, Advances in Chemical Physics, **51** (1982), pp. 1–48.

[18] S.E. Alm and S. Janson, *Random self-avoiding walks on one-dimensional lattices*, Commun. Statist.-Stochastic Models, 6 (1990), pp. 169–212.

[19] H.H. Schaefer, *Banach Lattices and Positive Operators*, (1974), (Berlin; New York : Springer-Verlag).

[20] C.E. Soteros, (1997), *in preparation*.

TOPOLOGICAL ENTANGLEMENT COMPLEXITY OF POLYMER CHAINS IN CONFINED GEOMETRIES

MARIA CARLA TESI*, E.J. JANSE VAN RENSBURG†, ENZO ORLANDINI‡, AND STUART G. WHITTINGTON§

Abstract. Long polymer chains in solution can be highly self- or mutually entangled. Several kinds of constraints can influence entanglement properties; among them a crucial role is played by geometrical constraints, which confine the chain(s) to restricted spaces. In this paper we investigate, using analytical and numerical techniques, the effects of different kinds of geometrical constraints on the topological entanglement complexity of polymer chain(s), i.e. on the knotting and linking probabilities.

Key words. Lattice models, self-avoiding polygons, topological entanglement complexity, knotting and linking probabilities, Monte Carlo methods.

1. Introduction. Linear polymer molecules in dilute solution are highly flexible and are characterized by a high degree of conformational freedom. As the molecules move in solution they can become both self-entangled and entangled with other molecules and these entanglements can have an important influence on a number of properties of the polymer. For instance these entanglements are important in determining the crystallization behaviour of polymers [1] and can influence rheological properties such as rubber elasticity [2]. If the polymers undergo a ring closure reaction these entanglements can be captured as knots and links in the resulting ring polymers and the original entanglements in the linear case can be investigated by studying the occurrence of knots and links in the resulting rings. In addition, analysis of the type of knots and links in circular DNA molecules can give important information on the mechanism of action of certain enzymes which break and reform DNA chains and rings [3,4].

Ring polymers (such as closed circular DNA) can be modelled as n-step self-avoiding polygons on a regular lattice, and the presence of knots in these polygons (and in related models, such as polygons in R^3) has been studied using Monte Carlo methods by a number of workers [5,6,7,8]. Little is known rigorously, but it has been shown that sufficiently long polygons are knotted with probability one [9,10]. These rigorous results have been successively extended to the more general case of graphs embedded in Z^3 [11] and to the problem of the entanglement complexity of self-avoiding walks [12].

* Université de Paris-Sud, Mathématiques Bat 425, 91405 Orsay Cedex, France. E-mail: Maria-Carla.Tesi@math.u-psud.fr.

† Department of Mathematics and Statistics, York University, Downsview, Ontario, Canada M3J 1P3. E-mail: rensburg@math.yorku.ca.

‡ CEA-Saclay, Service de Physique Théorique, F-91191 Gif-sur-Yvette Cedex, France. E-mail: enzo@spht.saclay.cea.fr.

§ Department of Chemistry, University of Toronto, Toronto, Ontario, Canada M5S 3H6. E-mail: swhittin@alchemy.chem.utoronto.ca.

In the case of the entanglement between different chains the simplest case to consider is linking of two ring polymers. The properties of linked pairs of polygons in these systems have been studied using Monte Carlo methods by a number of workers [6,13,14], whereas rigorously some work has been done on the linking probability for random embeddings of circles in R^3 [15,16] and for polygons on the cubic lattice [17].

Polymers are often confined to restricted spaces; for instance the presence of histones and other large molecules in the cell-nucleus confines the cellular DNA to reduced spaces, with corresponding effects on the geometrical and topological properties of the DNA. In this respect new questions can arise quite naturally concerning the effect of the geometrical constraints on the entanglement complexity of polymers. These questions will be addressed in this paper for ring polymers modelled by polygons in the cubic lattice, in the specific case where the polygons (or pairs of polygons for links) are confined to a slab or to a prism.

Section 2 gives a brief review of some pertinent results on self-avoiding walks and self-avoiding polygons in Z^3. In Section 3 we present some rigorous results on the knot probability for polygons; in particular after recalling previous results on the presence of random knots in unconstrained polygons, we investigate the large n-dependence of the knot probability for polygons confined to slabs and prisms of the simple cubic lattice. In Section 4 we extend the analysis of Section 3 to the linking probability for polygon pairs.

The results which will be discussed in Sections 3 and 4 are all asymptotic in nature. They say rather little about knotting and linking probabilities for small values of n. If one is interested in the behaviour for small n, there are at present no viable theoretical methods; the most useful current approaches are Monte Carlo simulation methods. In section 5 we explain the implementation of a Monte Carlo program to simulate polygons in the cubic lattice, and we explore the possibility of using rejection techniques to extract data on the properties of polygons confined to slabs and prisms. We present and discuss the results for the knot probability in slabs and prisms in Section 6 whereas in Section 7 we use similar methods to examine the behaviour of the linking probability.

The last section is devoted to summary and discussion.

2. Self-avoiding walks and polygons in the bulk. In order to understand most of the rigorous results which have been obtained about knotting and linking, one needs some background knowledge on self-avoiding walks. In this section we review some of the necessary results, and their relation to polygons. We shall be chiefly concerned with Z^3, the simple cubic lattice. The vertices of this lattice are the integer points in R^3, and the edges join pairs of vertices which are unit distance apart.

An *n-step walk* is an ordered sequence $\omega_0, \omega_1, \omega_2, \cdots, \omega_n$ of points of Z^3 such that successive points are unit distance apart; it is an *n-step self-avoiding walk* (which we shall abbreviate to an *n-SAW*) if these $n+1$ points are all distinct. We call the points $\omega_0, \omega_1, \cdots, \omega_n$ the vertices of the walk, and we consider two walks to be *equivalent* if one is a translate of the other. One is often interested in the properties of *n-SAWs* to within equivalence, and in this case it is convenient to consider the walks with the first vertex at the origin and to write c_n for the number of such walks.

An *n-step self-avoiding circuit* is an *(n-1)-SAW* whose first and last vertices ω_0 and ω_{n-1} are unit distance apart. If $\omega_0, \omega_1, \omega_2, \cdots, \omega_{n-1}$ are the successive vertices of an n-step self-avoiding circuit, then any cyclic permutation or reverse of a cyclic permutation of the vertices is also an n-step self-avoiding circuit. The equivalence class of $2n$ n-step self-avoiding circuits may be regarded as a single (unlabeled) geometrical object which we call an *n-step self-avoiding polygon* or simply an *n-SAP*. Two polygons are equivalent if one is a translate of the other, and we write p_n for the number of inequivalent *n-SAPS*. An *n-SAP* is said to be rooted if it contains one labeled vertex which coincides with the origin and, since the root can be chosen in n ways, the number of distinct rooted *n-SAPs* is np_n. In what follows, we shall use the term *polygon* to denote either *n-SAPs* or rooted *n-SAPs*. Hammersley [18] has shown that there exists a connective constant $\kappa > 0$ such that

$$p_n = e^{\kappa n + o(n)}. \tag{2.1}$$

The same result holds for walks, with the same connective constant κ [19].

3. Rigorous results on knot probabilities for polygons in confined geometries. Before discussing the knot probability in confined geometries a brief recall of the rigorous results for unconstrained random polygons is needed. We begin by giving a brief, non-technical account of a result which says that all except exponentially few sufficiently long polygons are knotted. The idea is conceptually very simple, and falls into two parts. The first is a pattern theorem due to Kesten [20]. A *pattern* is any finite self-avoiding walk. Kesten's theorem says that if there exists a self-avoiding walk on which a pattern K appears three times then the pattern will appear on almost all sufficiently long self-avoiding walks with positive density. We shall call such patterns *Kesten patterns.*

The second ingredient in the proof is the idea of knotted arc (see [9] for the technical definition of knotted arc). We can capture the important part of a knot such as the trefoil and tie it so tightly on the lattice that the rest of the walk cannot pass through the neighbourhood of this subwalk and untie the knot. Technically, the subwalk and its associated dual 3-cells form a knotted ball pair. These ball pairs can be constructed so that the subwalk is a Kesten pattern and hence the subwalk (and its associated dual 3-cells) must occur at least once on almost all sufficiently long self-avoiding walks. One can show that the pattern theorem also applies to polygons so that almost all sufficiently long polygons contain these knotted arcs and are therefore knotted. If we write p_n^o for the number of ***unknotted*** polygons, a concatenation argument shows that the limit $\lim_{n\to\infty} n^{-1}\log p_n^o$ exists so that

$$p_n^o = e^{\kappa_o n + o(n)}, \tag{3.1}$$

and Kesten's theorem implies that $0 < \kappa_o < \kappa$. In other words, the probability $P(n)$ that the polygon is a knot goes to unity exponentially rapidly as

$$P(n) = 1 - p_n^o/p_n = 1 - e^{-\alpha n + o(n)} \tag{3.2}$$

for some positive constant $\alpha = \kappa - \kappa_o$. One concludes that unknotted polygons comprise an exponentially small fraction of all polygons as n tends to infinity. In fact, the theorem is a good deal stronger. Since any knot can be associated with a Kesten pattern, every knot must appear somewhere along a sufficiently long polygon and, moreover, it must occur many times (in fact with positive density). This means that almost all the knots in long polygons are composite (i.e. they can be factorized into prime components) and the knotting (self entanglement complexity) is very bad [11]. Similar results are valid for various continuum versions of piecewise linear embeddings of circles in R^3 [21,22].

The results presented above hold for polygons in the bulk, i.e. for polygons in an unconstrained environment. Now we want to address similar questions in the case of polygons confined to certain subsets of the simple cubic lattice Z^3. We define an *L-slab* as the section graph of Z^3 whose vertex set is the set of vertices of Z^3 with z-coordinate in the range $0 \leq z \leq L$, and whose edge set is the set of edges of Z^3 incident on two of these vertices. Similarly an *(L_1, L_2)-prism* is the section graph of Z^3 with vertices having y and z coordinates satisfying $0 \leq y \leq L_1$ and $0 \leq z \leq L_2$, and with the edges of Z^3 which are incident on two of these vertices. Let $c_n(L)$ be the number of self-avoiding walks in an L-slab. Similarly, let $p_n(L)$ be the number of (unrooted) polygons confined to an L-slab. We regard two walks or two polygons confined to a slab as distinct if they can not be superimposed by translation in the x or y directions. We can now prove the analogous result to eqn.(3.2) for polygons confined to an L-slab. In order to do this, we recall the following lemma [23]:

LEMMA 1. ***[Hammersley and Whittington 1985]*** *If $c_n(L)$ is the number of n-step self-avoiding walks in an L-slab, the limit*

$$\lim_{n\to\infty} n^{-1}\log c_n(L) = \kappa(L) \tag{3.3}$$

exists for all L. Moreover $\kappa(L)$ is strictly monotone increasing in L and $\lim_{L\to\infty}\kappa(L) = \kappa$.

We next need to generalize the pattern theorem for walks to walks confined in an L-slab. For any $\alpha > 0$ and $\beta > 0$, define a $K_{\alpha,\beta}$ pattern to be any self-avoiding walk ω such that at least three disjoint copies of ω occur on some self-avoiding walk ω^*, where one endpoint of ω^* is the origin, and the other is $(\alpha, \beta, 0)$. In addition, ω^* is completely contained in $D_{\alpha,\beta} = \{(x, y, z) \in Z^3 : 0 \leq x \leq \alpha, 0 \leq y \leq \beta, 0 \leq z \leq \text{L}\}$. Then we have the following lemma:

LEMMA 2. *For any $\alpha > 0$ and $\beta > 0$, let P be a $K_{\alpha,\beta}$ pattern, then*

$$\limsup_{n\to\infty} n^{-1} \log c_n(\bar{P}, L) = \kappa(\bar{P}, L) < \kappa(L) \tag{3.4}$$

where $c_n(\bar{P}, L)$ is the number of n-step walks in an L-slab which do not contain the pattern P.

The proof of this lemma is similar to the case for walks in a prism [24], and we omit the details.

Walks and polygons have the same connective constant, and this is also true when they are confined to an L-slab; indeed the following lemma holds [25]:

LEMMA 3. ***[Madras and Slade 1993, Theorem 8.2.2]*** *If $p_n(L)$ is the number of unrooted polygons in an L-slab then*

$$\lim_{n\to\infty} n^{-1} \log p_n(L) = \kappa(L) \tag{3.5}$$

exists for all L. Moreover $\kappa(L)$ is strictly monotone increasing in L and $\lim_{L\to\infty} \kappa(L) = \kappa$.

Since deleting an edge cannot create a pattern we have

$$p_n(\bar{P}, L) \leq c_{n-1}(\bar{P}, L) \tag{3.6}$$

which, together with (3.4) and (3.5), establishes the following lemma:

LEMMA 4. *If there exists a self-avoiding walk in an L-slab on which a Kesten pattern P occurs then the number of polygons in an L-slab on which P never occurs is such that*

$$\limsup_{n\to\infty} n^{-1} \log p_n(\bar{P}, L) \leq \kappa(\bar{P}, L) < \kappa(L) \tag{3.7}$$

Lemma 4 implies that polygons in an L-slab which do not contain a given pattern P are exponentially rare compared to the total number of polygons in an L-slab.

As a consequence we have the following theorem [26], which extends the result in eqn. (3.2) to the case of self-avoiding polygons in an L-slab.

THEOREM 1. ***[Tesi et al. 1994]*** *If we denote by $p_n^0(L)$ the number of unknotted n-SAPs in an L-slab, then the knot probability $P(n, L)$ behaves as*

$$P(n,L) = 1 - p_n^0(L)/p_n(L) = 1 - \exp\left(-\alpha(L)n + o(n)\right) \tag{3.8}$$

with $\alpha(L) > 0$.

Proof: We take $T = \{-j, -j, k, -j, -j, -k, i, i, j, j, j, -i, k, -i, -i, -k, -j, -j, i, i, k\}$, where i, j, k are unit vectors in the coordinate directions, as the $K_{\alpha,\beta}$ pattern. This sequence of edges is a knotted arc in a 1-slab, and its presence in a polygon ensures that the polygon will be knotted. □

An immediate consequence is the following corollary:

COROLLARY 1. *All except exponentially few polygons in an L-slab in Z^3 are knotted.*

Some properties of two functions $\alpha(L)$ and $\kappa(L)$ are given in the following lemmas:

LEMMA 5. $\lim_{L\to\infty} \alpha(L) = \alpha$

Proof: We use equation (4.14) in [23] with $f(x) = L$, $N = L$ and $d = 3$ to prove that $\lim_{L\to\infty} \kappa(L) = \kappa$. Indeed from (4.14) in [23] we have

$$\liminf_{n\to\infty} n^{-1} \ln p_n(L) \geq L^{-1}(\ln p_L - 2\ln(2L^2)) \tag{3.9}$$

using (3.5) and letting $L \to \infty$ we obtain

$$\liminf_{L\to\infty} \kappa(L) \geq \kappa. \tag{3.10}$$

On the other hand, by inclusion we have $\kappa(L) \leq \kappa$, which implies

$$\limsup_{L\to\infty} \kappa(L) \leq \kappa, \tag{3.11}$$

so that

$$\lim_{L\to\infty} \kappa(L) = \kappa. \tag{3.12}$$

If we consider the number of unknotted unrooted polygons in an L-slab, $p_n^o(L)$, by concatenation techniques similar to those used to obtain (3.5), it is possible to show that the limit

$$\lim_{n\to\infty} n^{-1} \log p_n^o(L) = \kappa_o(L) \tag{3.13}$$

exists for all L. Moreover $\kappa_o(L) < \kappa(L)$ for all L. Since equation (3.9) can be generalized to unknotted polygons, and since by inclusion $\kappa_o(L) \leq \kappa_o$, we have

$$\lim_{L\to\infty} \kappa_o(L) = \kappa_o. \tag{3.14}$$

This gives the desired result, since $\lim_{L\to\infty} \alpha(L) = \lim_{L\to\infty}(\kappa(L) - \kappa_o(L)) = \kappa - \kappa_o = \alpha$. □

LEMMA 6. *$\kappa(L)$ and $\kappa_o(L)$ are concave functions of L.*

Proof: We consider two slabs of width L_1 and L_2, and we call $p_n^{(i,j)}(L_m)$ the number of n-gons in the slab of width L_m, with bottom edge distant i steps from the bottom face of the slab, and top edge distant j steps from the top face of the slab. Clearly $\sum_{i,j} p_n^{(i,j)}(L_m) = p_n(L_m)$. If we now consider the slab L_2 with bottom face coinciding with the top face of slab L_1, by concatenation of the polygons in these two slabs we obtain the following inequality:

$$p_n^{(i_1,j_1)}(L_1)\frac{p_n^{(i_2,j_2)}(L_2)}{2} \leq p_{2n+2(j_1+i_2-1)}^{(i_1,j_2)}(L_1+L_2). \tag{3.15}$$

By summing over all the indices in the above inequality we obtain:

$$\begin{aligned} p_n(L_1)\frac{p_n(L_2)}{2} &= \sum_{i_1,j_1,i_2,j_2} p_n^{(i_1,j_1)}(L_1)\frac{p_n^{(i_2,j_2)}(L_2)}{2} \\ &\leq \sum_{i_2,j_1}\sum_{i_1,j_2} p_{2n+2(j_1+i_2-1)}^{(i_1,j_2)}(L_1+L_2) \\ &= \sum_{i_2,j_1} p_{2n+2(j_1+i_2-1)}(L_1+L_2). \end{aligned} \tag{3.16}$$

Taking logarithms, dividing by $2n$ and letting $n \to \infty$ we obtain

$$\frac{1}{2}(\kappa(L_1)+\kappa(L_2)) \leq \kappa(L_1+L_2). \tag{3.17}$$

Exactly the same proof holds for $\kappa_o(L)$. □

To establish a result analogous to Theorem 1 for polygons confined to a prism the key ingredient one needs is a generalization of the pattern theorem for polygons in a prism. For technical reasons it is not possible to extend to the polygon case the techniques used for walks in [24]. However, using different techniques, a pattern theorem for polygons in a rectangular prism has been proved in [27]. From this one obtains an extension of Theorem 1 to the case of polygons in a prism, which implies that all except exponentially few sufficiently long polymers confined to a rectangular prism (of size (L, M), with $L \geq 2$ or $M \geq 2$) are knotted.

4. Rigorous results on linking probabilities of polygons pairs, unconstrained and subjected to constraints. The work described in the previous section is concerned with knotting as a measure of self-entanglement complexity. Now we want to address the problem of the entanglement complexity between several polygons focussing on the situation in which the polygons are confined to a slab or to a prism. We shall restrict ourselves to the simplest case of two linked polygons (links with 2 components). Two links L_1 and L_2 are *equivalent* if there is a homeomorphism of R^3 onto itself which takes L_1 to L_2, and the equivalence class of a link is called its *link type*.

There are three distinct definitions of linking used in topology, and we first discuss the relation between these. Two disjoint simple closed curves C_1 and C_2 are *topologically unlinked* if there is a homeomorphism of R^3 onto itself, $H : R^3 \to R^3$, such that the images $H(C_1)$ and $H(C_2)$ are separated by a two dimensional plane. The simple closed curve C_1 is *homotopically unlinked* from C_2 if there is a homotopy h_t from the embedding C_1 to the constant map (i.e. $h_0(C_1) = C_1$ and $h_1(C_1)$ is a point) such that $h_t(C_1)$ is disjoint from C_2 $\forall t \in [0,1]$. It is possible for C_1 to be homotopically unlinked from C_2 but for C_2 to be homotopically linked to C_1 so homotopic linking is not a symmetric relation [28]. Finally, C_1 is *homologically unlinked* from C_2 if C_1 bounds an orientable surface which is disjoint from C_2. Homological linking is a symmetric relation, and homological linking implies homotopic linking of each curve to the other which implies topological linking. In this paper we shall be mainly concerned with homological and topological linking.

It is easy to detect whether or not two curves are homologically linked [28]. One method, which is particularly useful for PL curves, is to orient each of the two curves C_1 and C_2, and to project them onto a plane so that no vertex in the projection of C_1 falls on the projection of C_2, or *vice versa*. At each point where C_1 crosses *under* C_2 we assign a value $+1$ or -1, according to the orientation of the crossing (see Figure 1). The sum of these crossing numbers is called the *linking number* of the two curves $l(C_1, C_2)$, and the curves are homologically linked if and only if $l(C_1, C_2) \neq 0$ [28]. Detecting topological linking is a more difficult problem, and the method we use here offers only a partial solution, in that it fails to detect some links. We compute the two variable Alexander polynomial $\Delta(s,t)$ evaluated at $s = t = -1$ (see for instance [29] for a justification of the method and [13] for details of the calculation). If $\Delta(-1,-1) \neq 0$ then the curves are topologically linked, but it is possible for a linked pair to have $\Delta(-1,-1) = 0$. However, this does not occur for any link with less than 9 crossings.

For linked polygon pairs not subjected to geometrical constraints, it has been proved [17] that given two polygons with fixed link type τ, such that both polygons are unknotted, other than the unlink, the connective constant of the polygon pair is equal to κ_o, independently of τ:

THEOREM 2. ***[Orlandini et al. 1994]*** *If $p_n^{(2)}(\tau)$ is the number of embeddings in Z^3, per lattice site, of two polygons, each with n edges, forming a link of type τ, where τ is any link with both circles unknotted, other than the unlink, then*

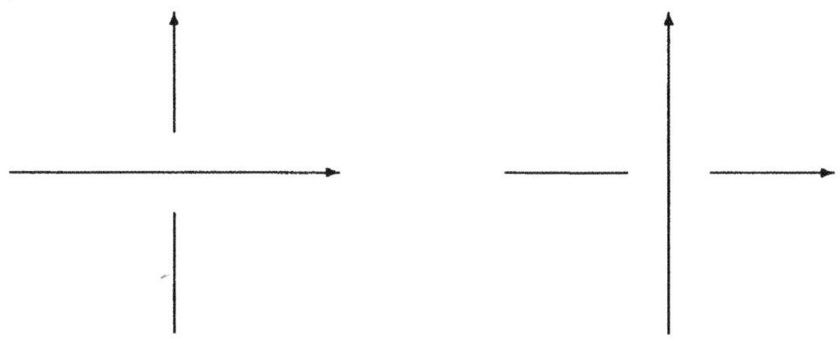

FIG. 1. *The sign convention for crossings*

(4.1)
$$\lim_{n\to\infty} \frac{\log p_n^{(2)}(\tau)}{2n} = \kappa_o.$$

It is interesting to compare what we know about this case compared to the case of polygons with fixed knot type σ. Writing $p_n(\sigma)$ for the number of n-gons with fixed knot type σ, the following inequality has been proved [11]:

(4.2)
$$\kappa_0 \le \liminf n^{-1} \ln p_n(\sigma) \le \limsup n^{-1} \ln p_n(\sigma) < \kappa.$$

Assuming that the limit $\kappa(\sigma) = \lim_{n\to\infty} n^{-1} \ln p_n(\sigma)$ exists, then (4.2) would imply that

(4.3)
$$\kappa_0 \le \kappa(\sigma) < \kappa.$$

I.e. the connective constant for polygons with fixed knot type would be different from κ and might be equal to κ_o. However, the existence of $\kappa(\sigma)$ has not been established for polygons with a fixed knot type.

We can also consider augmenting a link type τ by considering the union of all link types which can be obtained from τ by taking the connect sum of each circle with any knot (including the unknot). We call this union the *augmented link* $\tau^{\#}$.

Each of the polygons making up an augmented link (i.e. each component of the link) can be knotted, and one can ask for the probability that an embedding of an augmented link (of type $\tau^{\#}$) has unknotted components. The answer is given by the following theorem [17]:

THEOREM 3. *[**Orlandini et al. 1994**] The probability $P^{(2)}(n, \tau^{\#})$ that both components (with n edges) of an augmented link of type $\tau^{\#}$ are knotted goes to unity as*

$$P^{(2)}(n, \tau^{\#}) = 1 - e^{-\alpha n + o(n)} \tag{4.4}$$

when $n \to \infty$.

This result shows that each of the two components of the linked pair will be knotted in the $n \to \infty$ limit, except for exponentially few exceptions.

We show now how the results presented above can be extended to the class of polygons pairs confined to an L-slab. In particular, we can now prove the following theorem:

THEOREM 4. *If $p_n^{(2)}(L, \tau^{\#})$ is the number of embeddings in Z^3, per lattice site, of two polygons each with n edges, in a slab of width L, forming an augmented link of type $\tau^{\#}$, where τ is any link other than the unlink, then*

$$\lim_{n\to\infty} \frac{\log p_n^{(2)}(L, \tau^{\#})}{2n} = \kappa(L) \tag{4.5}$$

Proof: Define the *bottom vertex* of a polygon as the vertex having smallest coordinates, taken in lexicographic order.

To obtain an upper bound on $p_n^{(2)}(L, \tau^{\#})$, we note that the two polygons cannot be linked if the bottom vertex of one is not within a brick of volume $n^2 h$, where $h = \min\{L, n\}$ whose bottom vertex coincides with the bottom vertex of the other polygon. Hence we can construct each polygon in $p_n(L)$ ways, and translate one relative to the other in at most $n^2 h$ positions. Hence

$$p_n^{(2)}(L, \tau^{\#}) \leq n^3 p_n^2(L). \tag{4.6}$$

To obtain a lower bound we note that, for any given augmented link type $\tau^{\#}$, we can construct an embedding in an L-slab in Z^3 of a pair of polygons with link type τ. This follows by an extension of the arguments of [11]. By subdivision and concatenation we can arrange this embedding so that an edge e_1 of one polygon lies in the plane $x = x_1$, containing vertices with smallest x-coordinate, and an edge e_2 of the other polygon lies in the plane $x = x_2$ containing vertices with largest x-coordinate. Moreover, both polygons can be arranged to have the same number of edges, say $m = m(\tau)$. With this pair of polygons fixed, we can translate each of $p_{n-m}(L)/2$ polygons with $n - m$ edges so that the plane $x = x_2 + 1$ contains an edge of this polygon parallel to e_2, and no vertices of this polygon have x-coordinate less than $x_2 + 1$. These two polygons can now be concatenated by adding and deleting pairs of edges. The link type of the resulting polygons is the connected sum of the link τ and a knot (including the unknot), so the augmented link type does not change in this construction. The same construction can be carried out for the edge e_1, giving the bound

$$p_n^{(2)}(L, \tau^{\#}) \geq p_{n-m}^2(L)/4. \tag{4.7}$$

Taking logarithms, dividing by $2n$ and letting n go to infinity in (4.6) and (4.7) we obtain (4.5). □

Each of the polygons making up a link (i.e. each component of the link) can be knotted, and we can ask for the probability that an embedding of an augmented link (of type $\tau^{\#}$) has unknotted components. For free polygons it has been proved [17] that the probability that both components (with n edges) of a link of type $\tau^{\#}$ are knotted goes to unity exponentially rapidly when $n \to \infty$.

Also this theorem can be generalized to the case of linked polygons pairs confined to a slab, in the following way:

THEOREM 5. *The probability $P^{(2)}(n, L, \tau^{\#})$ that both components (with n edges) of an augmented link of type $\tau^{\#}$, confined to a slab of width L, are knotted goes to unity as*

$$P^{(2)}(n, L, \tau^{\#}) = 1 - e^{-\alpha(L)n+o(n)} \tag{4.8}$$

when $n \to \infty$.

Proof: Let $p_n^{(2)}(0, L, \tau^{\#})$ be the number of embeddings (per lattice site) of a pair of polygons, at least one of which is unknotted, each of which has n edges, confined to a slab of width L, and which are components of a non-trivial link of type $\tau^{\#}$. Then, by the argument used to obtain (4.6), but using unknotted embeddings of at least one of the polygons,

$$p_n^{(2)}(0, L, \tau^{\#}) \leq n^3(p_n^o(L))^2 + 2n^3 p_n^o(L) p_n(L). \tag{4.9}$$

Similarly,

$$p_n^{(2)}(0, L, \tau^{\#}) \geq p_{n-m}(L) p_{n-m}^o(L)/4. \tag{4.10}$$

Then

$$\begin{aligned} P^{(2)}(n, L, \tau^{\#}) &= 1 - p_n^{(2)}(0, L, \tau^{\#})/p_n^{(2)}(L, \tau^{\#}) \\ &\geq 1 - [n^3(p_n^o(L))^2 + 2n^3 p_n^o(L) p_n(L)]/[p_{n-m}^2(L)/4] \\ &= 1 - e^{-(\kappa(L)-\kappa_o(L))n+o(n)} - e^{-2(\kappa(L)-\kappa_o(L))n+o(n)} \\ &= 1 - e^{-(\kappa(L)-\kappa_o(L))n+o(n)}(1 + e^{-(\kappa(L)-\kappa_o(L))n+o(n)}) \end{aligned} \tag{4.11}$$

and

$$\begin{aligned} P^{(2)}(n, L, \tau^{\#}) &\leq 1 - [p_{n-m}(L) p_{n-m}^o(L)/4]/n^3 p_n^2(L) \\ &= 1 - e^{-(\kappa(L)-\kappa_o(L))n+o(n)} \end{aligned} \tag{4.12}$$

so that (4.8) follows with $\alpha(L) = \kappa(L) - \kappa_o(L)$. □

The same kind of analysis can be performed for polygons pairs in a (L_1, L_2)-prism obtaining the following result:

THEOREM 6. *If $p_n^{(2)}((L_1, L_2), \tau^\#)$ is the number of embeddings in Z^3, per lattice site, of two polygons each with n edges, in a prism of size (L_1, L_2), forming an augmented link of type $\tau^\#$, where τ is any link other than the unlink, then*

$$\lim_{n\to\infty} \frac{\log p_n^{(2)}((L_1, L_2), \tau^\#)}{2n} = \kappa(L_1, L_2) \tag{4.13}$$

Proof: The proof of this theorem is similar to the case for polygon pairs in an L-slab and we omit the details. The only difference consists in replacing Lemma 3 with the following lemma [24]:

LEMMA 7. ***[Soteros and Whittington 1989]*** *If $p_n(L_1, L_2)$ is the number of unrooted polygons in an (L_1, L_2)-prism then*

$$\lim_{n\to\infty} n^{-1} \log p_n(L_1, L_2) = \kappa(L_1, L_2) \tag{4.14}$$

exists for all pairs (L_1, L_2). □

5. Numerical methods. The results described in the last two sections are about asymptotic behaviour of polygons. If one is interested in the numerical values of properties such as the knotting probability or the linking probability at modest values of n, the only useful approach seems to be a Monte Carlo method. In addition this approach can be used to estimate parameters such as $\alpha(L)$.

The configurational and topological properties of polygons subject to geometrical constraints can be studied in two ways. First, one can use an efficient algorithm to generate a realization of a Markov chain defined on the set of n-gons and then use rejection techniques to select the subsequence of polygons in the realization which conforms to the constraints. The alternative approach is a direct simulation of polygons in the confined environment. Both these methods have disadvantages. In the first case, one may find that few of the polygons in the realization of the Markov chain satisfy the constraints, which may result in large uncertainties in any averages computed over the realization. In the second case, the problem is more basic: not much is known about Monte Carlo algorithms for polygons or walks in confined environments. In particular, there do not seem to be algorithms available which are known to be ergodic, though we give one example of an efficient and ergodic algorithm for polygons (in a 1-slab) in [26]. In both approaches, the essential requirements are the ergodicity of the algorithms and a high efficiency in sampling the relevant state space.

For most of the work reported here we have used the first approach, consisting in generating a realization of a Markov chain defined on the set of *all* n-gons, and using rejection techniques to compute averages over polygons satisfying the geometrical constraints. A modification of the "cut-and-paste" algorithm, introduced in [30] for the simulation of self-avoiding walks in the canonical ensemble, proved to be sufficient for this study. The algorithm was studied in detail in [31] where it was called the pivot algorithm. A version of the algorithm which proved effective for simulating polygons in the cubic lattice was invented in [32]. The necessary elementary moves for the algorithm include several elements of the octahedral group, the symmetry group of the cubic lattice. Two pivots are chosen uniformly on the polygon and a symmetry operation is carried out on the shorter of the two segments connecting the pivots, as in [33].

To investigate the behaviour of the link probability we have to generate pairs of polygons. We implemented a slight modification of the algorithm used in [33] to sample the space of single polygons. Two pivots are chosen uniformly on each polygon and a symmetry operation (one for each polygon) is carried out on one of the two segments connecting the pivots [17]. In addition to the usual self-avoidance condition for a single polygon, we require mutual avoidance between the two polygons; such requirements can be efficiently implemented by using hash-coding [34].

6. Numerical results: Knotting probability. Although the probability that a polygon of length n is knotted approaches unity exponentially rapidly with increasing n (Theorem 1), the knot probability is very small for intermediate values of n (up to about 2000 edges). It is therefore a considerable challenge to compute reliable values for the knot probabilities in confined spaces, where inefficient sampling by the algorithm may hamper reliable estimates.

Knots in polygons can be detected by computing the value of the Alexander polynomial $\Delta(t)$ at $t = -1$ [5,7]. If $|\Delta(-1)| \neq 1$ then the polygon is a knot. Otherwise we assume that it is the unknot. In fact the Alexander polynomial is not a perfect invariant, and is unable to distinguish every knot-type. For instance the prime knot 8_{11} has the same Alexander polynomial as the composite knot $3_1\#6_1$, and 8_{15} has the same Alexander polynomial as $3_1\#7_2$. In this paper we are not concerned with distinguishing pairs of knots but only in deciding if a knot is non-trivial, and no knot with 10 or less crossings has non-trivial Alexander polynomial. However, the first problem does occur for us at 10 crossings since we only calculate $|\Delta(-1)|$ and not the full polynomial, and there are two knots with 10 crossings which have $|\Delta(-1)| = 1$. Previous work [7] has shown that knots with 10 or more crossings are extremely rare so we are confident that the error made in neglecting knots with $|\Delta(-1)| = 1$ will be insignificant.

For every state sampled in the Markov chain we computed $\Delta(-1)$, the mean square radius of gyration R_n^2 and the spans in the three lattice directions, L_χ, where $\chi = x, y$ and z. Rejection techniques for computing averages depending on L were implemented by reading the subsequence of states with $L_\chi \leq L$ and determining weighted averages over the three directions x, y and z. This procedure worked well for moderate to large values of L, but produced increasingly uncertain data as L was decreased to lower and lower values. We were indeed interested in the limiting behaviour of the knot probability as $L \to 1$, but could not determine these values from our data. Consequently, we developed [26] an algorithm which was specifically designed for the case $L = 1$.

This algorithm was coded using the same techniques as set out above. Runs were performed under the same conditions as above and the same data were collected. In this case however, rejection techniques were not necessary in analyzing the data.

It is possible to prove the ergodicity of a cut-and-paste type algorithm in a $(1,1)$-prism, using exactly the arguments mentioned for the 1-slab. However, it is not necessary to compute the knot probability, since both intuitive and rigorous arguments show that no polygon in a $(1,1)$-prism can be knotted [26].

We ran the algorithm for mN attempted pivot transformations, where we determined the knot type of N polygons, with each observation separated by m attempted transformations. We chose $N = 50000$, with $m = 50$. We computed error bars on our estimates by calculating autocorrelation times in our MC simulations, using the techniques developed in [31].

6.1. Polygons confined in a L-slab. It is reasonable to expect an increase of the knot probability of polygons in an L-slab with decreasing L; the confined polygons become smaller in size and are more likely to be knotted [7]. However, it is not at all clear whether this trend will continue as L tends to 1. In Figure 2 we show the knot probability as a function of L for polygons of lengths n=600, 800, 1200 and 1600 confined in slabs of different widths.

For each fixed n the probability appears to reach a maximum for a finite value of $L = L(n)$ and the fraction of knotted polygons decreases with L if $L < L(n)$. To confirm the existence of the maxima at $L = L(n)$, we computed the knot probability at $L = 1$ using the 1-slab pivot algorithm as set out above. We found a knot probability which is almost zero for every n. The presence of a maximum in the probability may reflect the competition between the higher probability for a polygon with smaller radius of gyration to be knotted, and the difficulty of forming knots in confined geometries due to the steric interactions in the polygon. If the polygon is squeezed along a given axis to a width L, then it spreads in the other directions, thus increasing the mean square radius of gyration again, and one expects a decrease in the fraction of knotted conformations. Eqn (3.8) suggests that the fraction of unknotted conformations in an L-slab, $P^o(n, L)$, behaves as

$$P^o(n, L) \simeq \mathcal{A} e^{-\alpha(L)n} \tag{6.1}$$

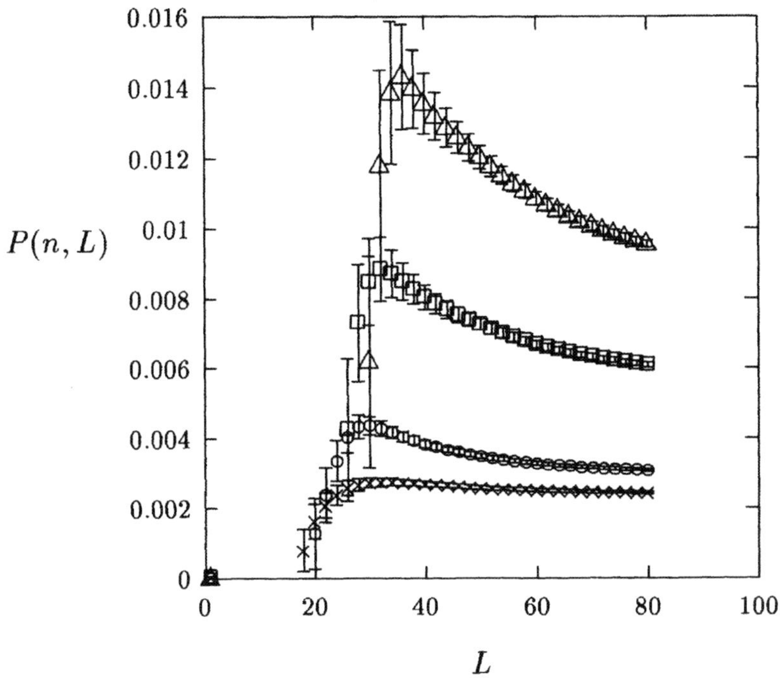

FIG. 2. *Fig. 1 Plot of the knot probability $P(n, L)$ versus L for polygons of different sizes n ($n = 600(\times), n = 800(\circ), n = 1200(\square), n = 1600(\triangle)$) confined in slabs of width L.*

In Figure 3 we plot $\ln P^o(n, L)$ as a function of n for $L = 40, 50, 60$. The straight lines shown are the fits obtained from a weighted least square linear regression. This linear behaviour supports (6.1); moreover it is clear that the value of $\alpha(L)$ increases as L decreases. If we take the error bars to be three standard deviations, and if we discard points corresponding to polygons of small size, i.e. with $n = 600$, we obtain the following values for α: $\alpha(L = 40) = (1.23 \pm 0.20) \times 10^{-5}$, $\alpha(L = 50) = (1.07 \pm 0.16) \times 10^{-5}$ and $\alpha(L = 60) = (9.5 \pm 1.4) \times 10^{-6}$.

In Figure 4 we plot the mean-square radius of gyration against L for different values of n. For fixed n, each curve has a minimum for approximately the same $L = L(n)$ for which the knot probability reaches its maximum. This observation supports the idea that knottedness in polygons is closely correlated with the "size" of the conformation: the more compact the polygon, the more likely it will be knotted. Moreover, the polygons in the 1-slab have very large mean square radii of gyration, and we observe a low knot probability.

6.2. Polygons confined in an (L, L)-prism. We performed the same kind of analysis for polygons in an (L, L)-prism. In this case the polygon fits into the prism if L_a and L_b are both less than L, with $a = x, b = y$,

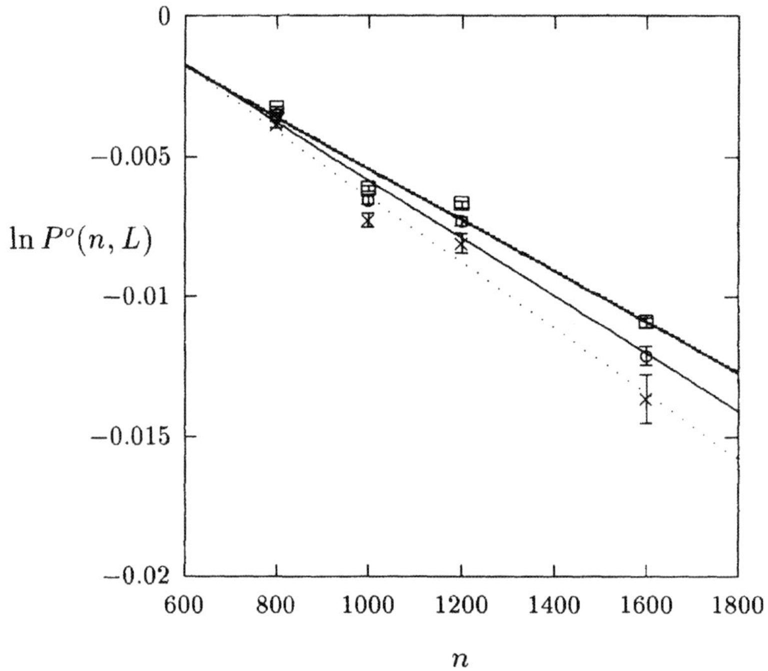

FIG. 3. *Plot of* $\ln P^o(n,L)$, *for a polygon in a slab, against n for three fixed values of* L ($L = 40(\times), L = 50(\circ), L = 60(\square)$). *The straight lines are least squares fits.*

or $a = x, b = z$, or $a = y, b = z$. Weighted averages in three directions were taken to compute means over the ensemble of polygons by rejection techniques.

In Figure 5 we plot the knot probability as a function of L for different values of n.

We see exactly the same kind of behaviour found in the case of the slab. The only difference is in the values of the maxima, which are considerably higher than in the slab case; in a prism a polygon is more confined, and therefore more compact, and we expect a higher probability of being knotted. This is confirmed also from weighted linear fits of the quantities $\ln P^o(n, L)$ as a function of n for three different fixed values of $L(L = 40, 50, 60)$. There is a clear dependence of α on L: $\alpha(L = 40) = (2.25 \pm 0.30) \times 10^{-5}$, $\alpha(L = 50) = (1.70 \pm 0.15) \times 10^{-5}$ and $\alpha(L = 60) = (1.29 \pm 0.15) \times 10^{-5}$. The mean square radius of gyration exhibits a minimum at approximately the value of L where the knot probability has a maximum.

7. Numerical results: Linking probability. We are interested in studying situations in which the two polygons cannot be too far apart in space, since otherwise they will be unlinked. To focus the sampling on this

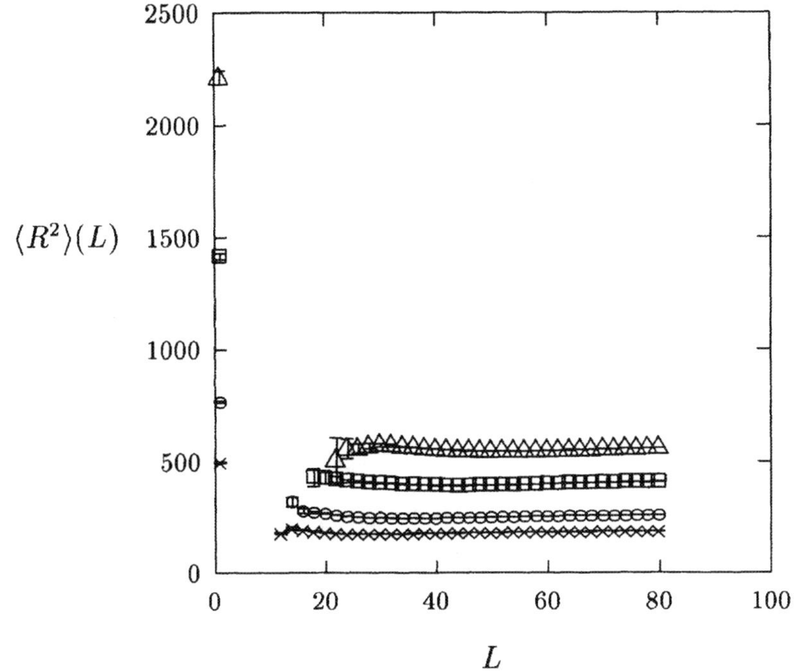

FIG. 4. *Plot of* $\langle R_n^2\rangle(L)$ *versus* L *for different values of* n *(*$n = 600(\times), n = 800(\circ), n = 1200(\Box), n = 1600(\triangle)$*).*

regime we use an importance sampling method where polygons are sampled from the probability distribution function $f(d, d_0) = A\exp(-d^2/d_0^2)$ where d is the distance between the centres of mass of the two polygons, d_0 is an appropriately chosen parameter, and A is a normalization constant. Since we are sampling from a non-uniform distribution, in the data analysis we attach a weight w to each polygon pair in the sample, with $w = 1/f = A^{-1}\exp(d^2/d_0^2)$. In addition we pool the data from runs with different values of d_0 by using suitably weighted averages.

We studied polygon pairs of length n ranging from $n = 400$ up to $n = 1800$. For a fixed value of n we sampled with different values of d_0 in the interval $[n/4, 4n]$. For every state sampled in the Markov chain we computed the linking number, the Alexander polynomial evaluated at $s = t = -1$, $\Delta(-1,-1)$, the Alexander polynomial of each of the two components of the link, the distance d between the centres of mass of the two polygons, the mean square radius of gyration of each polygon, and the spans in the three lattice directions, L_χ, where $\chi = x, y$ and z, of the pair of polygons. Averages were taken over 25000 samples, sampled every 50 attempted pivots.

To investigate the dependence of the link probability on a geometrical constraint, such as a cube of side L, we used rejection techniques to choose,

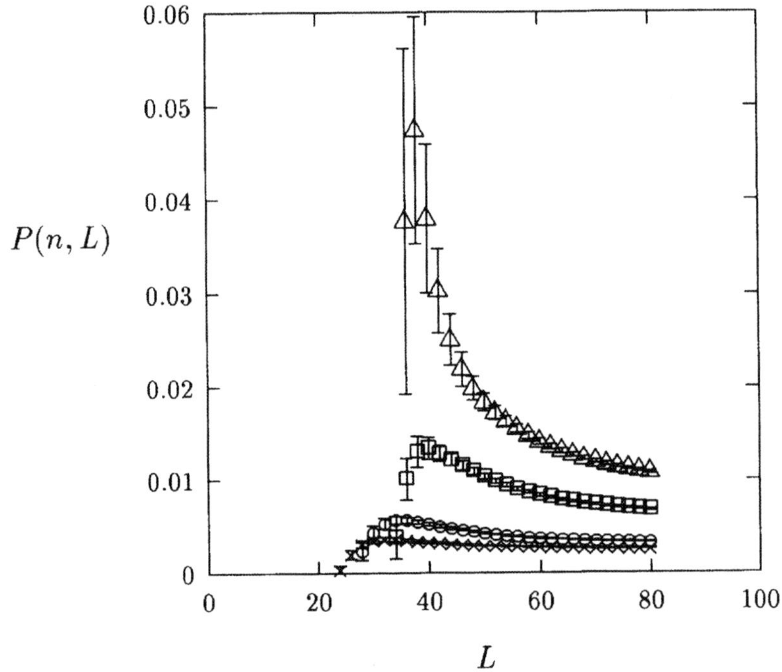

FIG. 5. *Plot of the knot probability $P(n, L)$ versus L for polygons of different sizes n ($n = 600(\times), n = 800(\circ), n = 1200(\square), n = 1600(\triangle)$) confined in prisms of width (L, L).*

for a fixed L, the subset of pairs of polygons which could be contained in the L-cube. A pair of polygons fits a given L-cube if all the three spans are less than L. To count the number of ways in which a given pair can be contained in an L-cube we consider every two embeddings of pairs to be distinct even if they can be superimposed by translation in the x, y or z directions within the cube.

For the study of the link probability for polygons confined to an L-slab we used the same rejection techniques, sampling from the previously described distribution $f(d, d_0)$, with the parameter d_0 appropriately chosen.

7.1. Polygons pairs confined to cubes. The results obtained in section 4, especially Theorem 2, were for pairs of polygons whose configurations were not seriously affected by the presence of a constraint. For two polygons each with n edges, confined to a cube of side L, we would expect that this behaviour would continue until $L \sim n^\nu$ where ν is the exponent characterizing the n-dependence of the mean-square radius of gyration. For $L < n^\nu$, the polygons will be more compact, and this effect will increase as L decreases at fixed n. We shall focus on this regime.

In Figure 6 we show the probability for polygon pairs to be homologically linked (i.e. of having linking number non-zero) as a function of n and L.

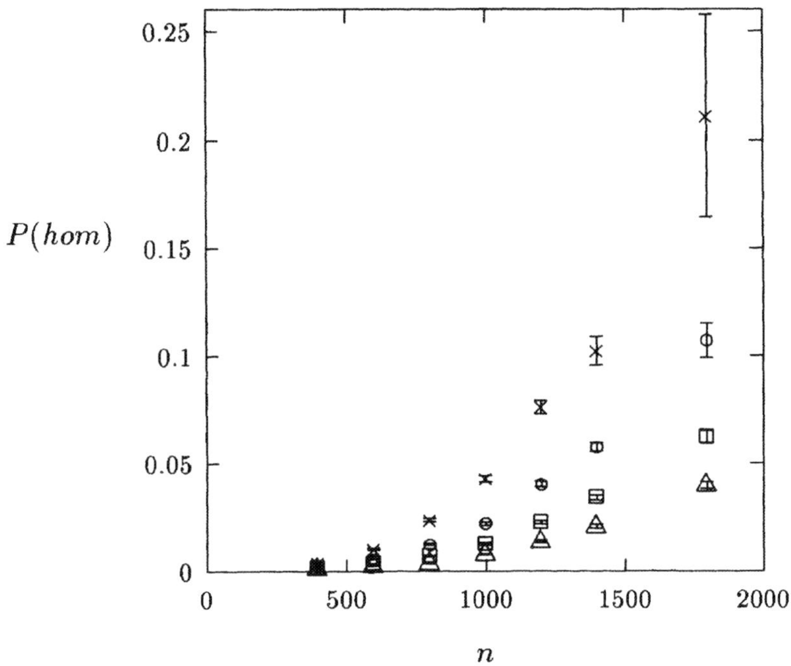

FIG. 6. *Probability for polygon pairs to be homologically linked (i.e. of having linking number non-zero) as a function of* n, *for several fixed values of* L ($L = 72(\times), L = 84(\circ), L = 96(\square), L = 108(\triangle)$).

The qualitative features are clear. The linking probability increases with increasing n at fixed L, and with decreasing L at fixed n. Since there are two length scales in this problem (n^ν and L) one expects that it will be the *ratio* between these length scales which will govern the behaviour, and this is indeed the case. In Figure 7 we show the linking probability as a function of $n/L^{1/\nu}$, and it is clear that all the data for different values of n and L fall on a single curve. We note that the linking probability is still quite small for $n/L^{1/\nu} = 1$.

Since homological linking is the weakest form of linking, it is interesting to investigate how effective it is as an indicator of topological linking. We consider the subset of pairs of polygons which have either $\Delta(-1,-1) \neq 0$ or $l \neq 0$ (or both), and we refer to these as topologically linked. (Although these invariants may miss some topologically linked pairs we do not expect that this will be quantitatively serious at these values of n.) In Figure 8 we plot the probability for a polygon pair of being homologically linked, given that the polygons are topologically linked, as a function of $n/L^{1/\nu}$. As we see in Figure 8 there are two different regimes corresponding to the dominance of one of the two length scales over the other. In particular for $n << L^{1/\nu}$, which corresponds to large boxes, almost all the topologically

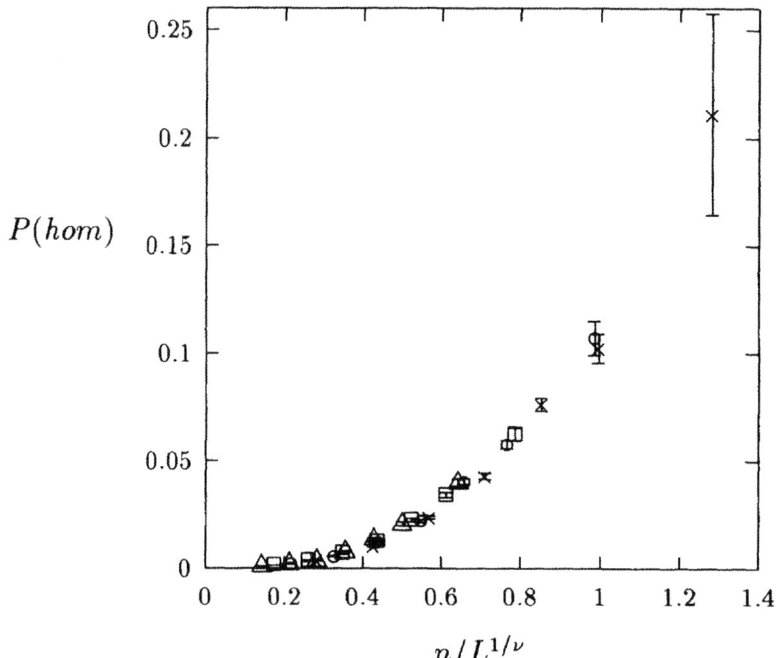

FIG. 7. *Linking probability as a function of* $n/L^{1/\nu}$*: all the data for different values of* n *and* L*, the same as in Figure 7.1, fall on a single curve.*

linked pairs are homologically unlinked. On the other hand, as $n/L^{1/\nu}$ increases, the fraction of topologically linked polygon pairs that are also homologically linked increases and goes to unity for $n >> L^{1/\nu}$. This behaviour, which is not obvious *a priori*, suggests that homological linking is a good indicator of topological linking for configurations in which the two polygons are strongly interpenetrating whereas it becomes a less efficient indicator for configurations in which the two components are, on average, further apart.

7.2. Polygons pairs confined to slabs. In this section we consider pairs of polygons (each with n edges) with distance between their centres of mass no greater than d. Once again we have two length scales (d and n^ν) and we expect that it will be the ratio of these length scales which will be important. Also in this case we can see the presence of two different regimes. For $n << d^{1/\nu}$ the centres of mass of the two polygons are very distant apart and the linking probability is almost zero. As $n/d^{1/\nu}$ increases, the two polygons come closer together and are more likely to be linked.

To evaluate the link probability for polygons pairs confined to an L-slab the usual rejection techniques have been used. The global qualitative

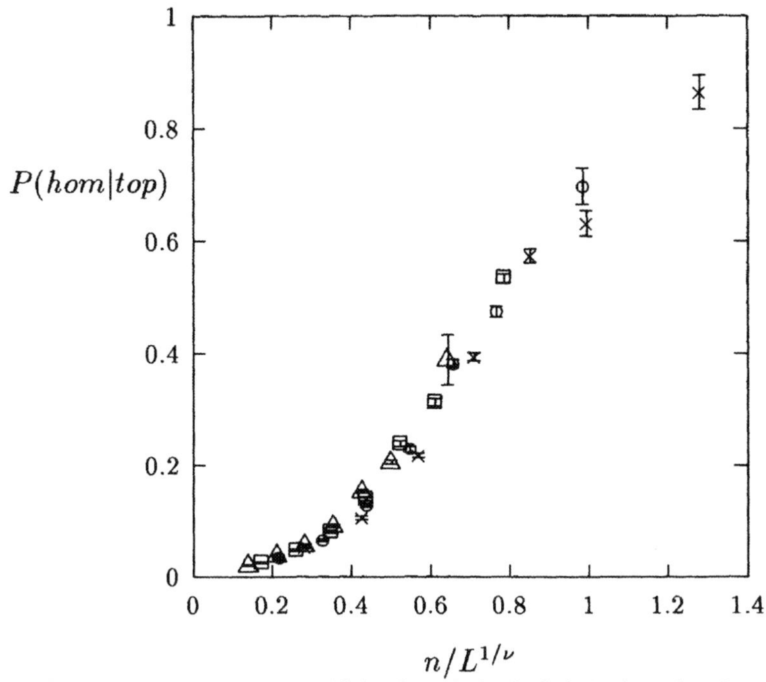

FIG. 8. *Probability of polygon pairs of being homologically linked, given that the polygons are topologically linked, as a function of* $n/L^{1/\nu}$.

features we have observed for the linking probability are very similar to the ones observed for the knotting probability. In particular also in the case of a polygon pair, confined to a slab or to a prism, the linking probability goes through a maximum when plotted as a function of L, the location of which depends on n. This indicates that the knotting and linking probabilities of ring polymers are influenced essentially in the same way by the presence of geometrical constraints such as slabs or prisms, suggesting that the two quantities are indicators of essentially similar topological features.

8. Discussion. In this paper we have studied the influence of geometrical constraints (such as slabs and prisms) on the knotting and linking probabilities of ring polymers.

We have shown rigorously that the knotting probability of all except exponentially few polymers in a slab or prism goes to one exponentially rapidly as $n \to \infty$. We have also shown that both components of a link are knotted with high probability. We have characterized the L dependence of the exponential factor $\alpha(L)$, showing that $\lim_{L\to\infty} \alpha(L) = \alpha$.

The small n behaviour has been investigated numerically, by Monte Carlo techniques, and similar features have been observed for the knotting and linking probabilities. More precisely both of them show a maximum when plotted as functions of L (in L-slabs or (L, L)-prisms) for fixed values of n. This may reflect the competition between the higher probability for constrained rings to be entangled (knotted or linked or both), and the difficulty of forming knots and links in confined geometries due to the steric interaction in the chain(s).

Of course, there are still interesting partially open questions, concerning for example the reciprocal influence of geometrical constraints and of the topological structure of the polymers, other than linear chains or polygons [27], or the influence of the quality of the solvent on knotting [26] and linking. Among others questions it would be of particular interest to have information on less regular constraints than the ones considered here.

Acknowledgements. The authors wish to thank Chris Soteros and DeWitt Sumners for many fruitful discussions. Part of this work was carried out while the authors were visitors at the IMA, and they would like to thank the IMA for hospitality and financial support. This work was partially supported by NSERC of Canada.

REFERENCES

[1] P.G. de Gennes, *Tight knots*, Macromolecules **17** (1984), 703–704.

[2] S.F. Edwards, *Statistical mechanics with topological constraints: I*, Proc. Phys. Soc. **91** (1967), 513–519.

[3] S.A. Wasserman and N.R. Cozzarelli, *Biochemical topology: applications to DNA recombination and replication*, Science **232** (1986), 951–960.

[4] D.W. Sumners, *Knot theory and DNA*, in *New Scientific Applications of Geometry and Topology*, Proceedings of Symposia in Applied Mathematics **45**, D.W. Sumners, ed., Am. Math. Soc. (1992), 39–72.

[5] A.V. Vologodskii, A.V. Lukashin, M.D. Frank-Kamenetskii and V.V. Anshelevich, *The knot probability in statistical mechanics of polymer chains*, Sov. Phys.-JETP **39** (1974), 1059–1063.

[6] J.P.J. Michels and F.W. Wiegel, *On the topology of a polymer ring*, Proc. Roy. Soc. A **403** (1986), 269–284.

[7] E.J. Janse van Rensburg and S.G. Whittington, *The knot probability in lattice polygons*, J. Phys. A: Math. Gen. **23** (1990), 3573–3590.

[8] K. Koniaris and M. Muthukumar, *Self-entanglement in ring polymers*, J. Chem. Phys. **95** (1991), 2873–81.

[9] D.W. Sumners and S.G. Whittington, *Knots in self-avoiding walks*, J. Phys. A: Math. Gen. **21** (1988), 1689–1694.

[10] N. Pippenger, *Knots in random walks*, Disc. Appl. Math. **25** (1989), 273–278.

[11] C.E. Soteros, D.W. Sumners and S.G. Whittington, *Entanglement complexity of graphs in* Z^3, Math. Proc. Camb. Phil. Soc. **111** (1992), 75–91.

[12] E.J. Janse van Rensburg, D.W. Sumners, E. Wasserman and S.G. Whittington, *Entanglement complexity of self-avoiding walks*, J. Phys. A: Math. Gen. **25** (1992), 6557–66.

[13] A.V. Vologodskii, A.V. Lukashin and M.D. Frank-Kamenetskii, *Topological interactions between polymer chains*, Sov. Phys.-JETP **40** (1974), 932–936.

[14] K.V. Klenin, A.V. Vologodskii, V.V. Anshelevich, A.M. Dykhne and M.D. Frank-Kamenetskii, *Effect of excluded volume on topological properties of circular DNA*, J. Biomol. Str. Dyn. **5** (1988), 1173–85.

[15] W.F. Pohl, *The probability of linking of random closed curves*, Lectures Notes in Mathematics **894** (1981), 113–126.

[16] B. DUPLANTIER, *Linking Numbers, Contacts and Mutual Inductances of a Random Set of Closed Curves*, Comm. Math. Phys. **82** (1981), 41–68.

[17] E. ORLANDINI, E.J. JANSE VAN RENSBURG, M.C. TESI AND S.G. WHITTINGTON, *Random linking of lattice polygons*, J. Phys. A: Math. Gen. **27** (1994), 335–45.

[18] J.M. HAMMERSLEY, *The number of polygons on a lattice*, Proc. Camb. Phil. Soc. **57** (1961), 516–523.

[19] J.M. HAMMERSLEY, *Percolation processes II. The connective constant*, Proc. Camb. Phil. Soc. **53** (1957), 642–645.

[20] H. KESTEN, *On the number of self-avoiding walks*, J. Math. Phys. **4** (1963), 960–969.

[21] Y. DIAO, N. PIPPENGER AND D.W. SUMNERS, *On random knots*, J. Knot Theory and Ramifications **3** (1994), 419–429.

[22] Y. DIAO, *The knotting of equilateral polygons in R^3*, J. Knot Theory and Ramifications **2** (1994), 413–425.

[23] J.M. HAMMERSLEY AND S.G. WHITTINGTON, *Self-avoiding walks in wedges*, J. Phys. A: Math. Gen. **18** (1985), 101–111.

[24] C.E. SOTEROS AND S.G. WHITTINGTON, *Lattice models of branched polymers: effects of geometrical constraints* , J. Phys. A: Math. Gen. **22** (1989), 5259–5270.

[25] N. MADRAS AND G. SLADE, *The Self-Avoiding Walk*, Birkhäuser, Boston, 1993.

[26] M.C. TESI, E.J. JANSE VAN RENSBURG, E. ORLANDINI AND S.G. WHITTINGTON, *Knot probability for lattice polygons in confined geometries*, J. Phys. A: Math. Gen. **27** (1994), 347–60.

[27] C.E. SOTEROS, *Knots in graphs in subsets of Z^3*, in this IMA volume.

[28] D. ROLFSEN, *Knots and Links*, Mathematics Lecture Series 7, Publish or Perish, Inc., Houston, Texas, 1976.

[29] G. TORRES, *On the Alexander polynomial*, Ann. Math. **57** (1953), 57–89.

[30] M. LAL, *"Monte Carlo" computer simulations of chain molecules. I.*, Molec. Phys. **17** (1969), 57–64.

[31] N. MADRAS AND A.D. SOKAL, *The pivot algorithm: A highly efficient Monte Carlo method for the self-avoiding walk*, **50**, J. Stat. Phys. (1988), 109–186.

[32] N. MADRAS, A. ORLITSKY AND L.A. SHEPP, *Monte Carlo generation of self-avoiding walks with fixed endpoints and fixed length*, J. Stat. Phys. **58** (1990), 159–183.

[33] E.J. JANSE VAN RENSBURG, S.G. WHITTINGTON AND N. MADRAS, *The pivot algorithm and polygons*, J. Phys. A: Math. Gen. **23** (1990), 1589–1612.

[34] D.E. KNUTH, *The Art of Computer Programming*, volume 3, Addison-Wesley, Reading MA, 1973.

SURVEY OF SELF-AVOIDING RANDOM SURFACES ON CUBIC LATTICES: ISSUES, CONTROVERSIES, AND RESULTS*

T.L. EINSTEIN† AND A.L. STELLA‡

1. Introduction and motivation. The aim of this short review is to present an overview of problems involving random surfaces on a cubic lattice: why one studies them, what questions one tries to answer, what assumptions are made in the models, what one computes to understand these models, and some things that have been learned. The emphasis will be on what one can calculate and why the problem is computationally intensive but tractable. There has also been considerable progress in rigorous treatments of this subject; this train of work will be discussed in the following paper. Since there have been some rather extensive reviews [1,2,3,4] on the subject, there is no need to be encyclopedic. Liberal use is made of these sources as well as unpublished lecture notes by A. Maritan [5] and E. Orlandini [6].

Fröhlich [1] lists a broad range of applications for the seemingly abstract random-surface problems. In quantum field theory they serve as representations of lattice guage theories and approximations to gauge theory. They can be used as a lattice approximation to string theory; i.e. they amount to the world sheets of strings in (3+1)D. There are also applications to 2D quantum gravity.

In condensed matter, they can be used to model crystal growth, crystal surfaces, interfaces and domain walls between phases (e.g. in fluids and magnets or in the problems of roughening and wetting), domain walls and their wandering in uniaxial commensurate-incommensurate transitions, and fluctuations of domain walls bounded by frustration loops in spin-glass models with short-range interactions, to calculate the contributions to free energy and entropy. In catalytic chemistry they can provide information about reactions on surfaces with "large" Hausdorff dimension d_H.

Arguably their most important applications are in the newly popular (cf. the recent text by P.M. Chaikin and T.C. Lubensky [9] as well as a host of monographs) field of "soft" condensed matter. This field encompasses the physics of polymers, vesicles, membranes, interfaces, aggregates, emulsions, and proteins. Such systems are characterized by a state lying

---

* Based on work in collaboration with E. Orlandini, I. Beichl, F. Sullivan, M.C. Tesi, J. O'Connell, D. Libes, U. Glaus, A. Maritan, and J.R. Banavar.

† Department of Physics, University of Maryland, College Park, MD 20742-4111, USA.

‡ INFM-Dipartimento di Fisica e Sezione INFN dell' Università di Padova, 35131 Padova, Italy.

between solid and liquid that has great flexibility and large fluctuations, but also strong topological constraints. In particular, random surfaces are excellent models for studying the statistical mechanics of branched and membrane-like polymers, of microemulsions, etc. The phase transitions of random surfaces are expected to include many different universality classes. As in other soft condensed matter systems, the symmetry is usually not evident from Hamiltonian; changes in class can result from changes in the interactions with no apparent change in symmetry (cf. polymer collapse) [4].

2. Models and definitions. We work with the integer points in R^3, i.e. a cubic lattice Z^3 (which can be generalized to a hypercubic lattice Z^d), with *edges* joining points separated by unit distance. A *plaquette* p is a unit square in Z^3. Two plaquettes are *joined* when they share the same edge ℓ and are *connected* when they belong to a sequence of plaquettes in which neighboring pairs are joined. A *random surface* S is a set of such connected plaquettes. It is *self-avoiding* when each plaquette occurs only once in the surface and at most two plaquettes meet at any edge. An illustrative example is provided in Fig. 1. There is no restriction, however, on how many plaquettes may meet at a particular point on Z^3, as indicated by the lower right corner of the top center cube of Fig. 1. [Without the self-avoidance restriction, the model is called planar random surfaces [1].] The connection between this model and lattice gauge theories was fully elucidated a decade ago [7]. In analogy with self-avoiding walks [8], the statistics of a single self-avoiding surface is recovered from a lattice gauge model (a spin model in the polymer case) as the number of components of some spin degrees of freedom approaches zero. We are particularly interested in the case when the surface is *closed*: every edge on the surface joins two plaquettes and there is no boundary (i.e. it is a fixed null boundary). Such structures are called vesicles [10]. In most studies, these vesicles are restricted to genus zero (so homeomorphic to a sphere), but we shall discuss the effect of higher genus, i.e. allowing handles. In addition, one can introduce 1) an *osmotic pressure*, a difference Δp of the interior minus the exterior pressures and 2) a bending rigidity, in this lattice structure an extra energy κ when two plaquettes join at right angles (rather than coplanar).

Let S denote a given surface configuration. As in the case of random walks, it is useful to determine the number of surface configurations (per lattice site) $N(n)$ with $|S| = n$ plaquettes and the relative average radius of gyration

$$R_n^2 = n^{-2} \sum_{|S|=n} \sum_{p \in S} \sum_{p' \in S} |r(p) - r(p')|^2 / N(n), \tag{2.1}$$

where $r(p)$ denotes the center of plaquette p of S. We then expect that asymptotically

$$R_n^2 \underset{n \to \infty}{\sim} n^{2\nu} \qquad \text{and} \qquad N(n) \underset{n \to \infty}{\sim} n^{-\theta} \mu^{-n}. \tag{2.2}$$

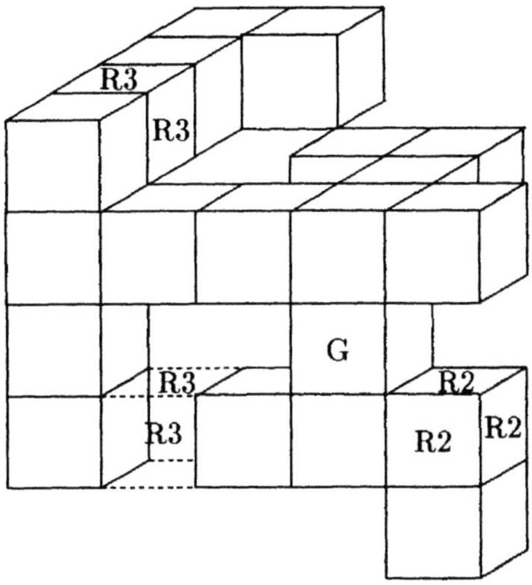

FIG. 1. *Illustration of a typical self-avoiding random surface, extended from Fig. 2 of ref. [2]. Its features are discussed in section II. The annotations R2 and R3 refer to forbidden types of moves in the Sterling-Greensite [17] formulation, as discussed in section III.*

Here ν is the inverse of the fractal dimension of S and θ is related to its entropy. Regarding the entropic interpretation, Durhuus *et al.* [11] showed that the large-n limit of $n^{-1} \ln N(n)$ exists and is finite. (This result was proved as an extension of a theorem for self-avoiding walks [12].) While the canonical picture of eqns. (2.1) and (2.2) has recently been used extensively by Baumgärtner [13,14,15], most work has relied on a grand-canonical ensemble. If we imagine a surface energy γ associated with each plaquette, then the probability of occurrence of a configuration in equilibrium should be proportional to $\exp(-\gamma|S|/k_BT)$, where k_BT is the product of Boltzmann's constant and the absolute temperature. Since the argument of the exponential depends only on the the dimensionless ratio γ/k_BT, it is simpler to deal with a plaquette fugacity $K \equiv \exp(-\gamma/k_BT)$. The grand partition function, or equivalently the generating function G, can in this simplest case be written

$$G(K) = \sum_S K^{|S|}, \tag{2.3}$$

where there is no restriction on $|S|$ in the summation. More generally, we shall allow for a pressure fugacity $w = \exp(\Delta p/k_BT)$ associated with osmotic pressure and another associated with bending rigidity $\rho = \exp(-\kappa/k_BT)$. We also include a weighting factor n^q for reasons which will become evident in eqn. (2.8). The generating function then becomes

$$(2.4)\quad G_q(K,\rho,w) = \sum_S |S|^q K^{|S|} \rho^{\ell(S)} w^{V(S)} = \sum_n n^q K^n Z_n(\rho,w),$$

where $|S|$, $\ell(S)$, and $V(S)$ denote the area, the effective number of edge links (actually the number of plaquettes meeting at right angles, which may exceed the physical number of links when "R2" moves are allowed, as discussed in Section V), and the volume enclosed by S, respectively, and $Z_n(\rho,w)$ is the canonical partition function for vesicles of area n.

To characterize the properties of the vesicles, we consider the limit $n \to \infty$. For example, in a critical regime, $Z_n(\rho,w)$ should diverge asymptotically as

$$(2.5)\qquad Z_n(\rho,w) = \sum_{|S|=n} \rho^{\ell(S)} w^{V(S)} \underset{n\to\infty}{\sim} n^{-\theta}[K_c(\rho,w)]^{-n},$$

where, as in eqn. (2.2), θ is the entropic exponent. (To simplify notation, we indicate the critical fugacity by K_c for all our cases, regardless of its changing arguments.) For $\rho = 1, w = 1$, Z_n is the number of distinct vesicle configurations with n plaquettes. Evaluation of θ by directly fitting Monte Carlo data with eqn. (2.5) does not work well. A much better strategy is to extract θ from the behavior of the average area of the vesicle plotted as a function of K [16]. In this grand-canonical ensemble, the average of some generic quantity $A(|S|, V(S), \ell(S))$ is

$$(2.6)\quad \langle A\rangle_q = G_q(K,\rho,w)^{-1} \sum_{|S|} A(|S|, V(S), \ell(S))|S|^q K^{|S|} \rho^{\ell(S)} w^{V(S)}.$$

For the special case of $A = |S|$, we see that

$$(2.7)\qquad \langle |S|\rangle_q = K\frac{\partial}{\partial K} \ln G_q(K,\rho,w).$$

Using the asymptotic form of $N(n)$ and taking the continuum limit, we find that

$$(2.8)\qquad G_q(K,\rho,w) \sim \frac{\Gamma(1+q-\theta)}{[\ln(1/\mu K)]^{1+q-\theta}}.$$

Notice that this expression is divergent only if $1+q-\theta$ is positive, so q must be chosen greater than $\theta - 1$ in order for singular behavior to be easily seen. Identifying $K_c(\rho,w)$ as μ^{-1} and inserting eqn. (2.8) into eqn. (2.7), we find

$$(2.9)\qquad \langle |S|\rangle_q \underset{K\to K_c^-}{\sim} (1+q-\theta)\frac{K_c(\rho,w)}{K_c(\rho,w)-K}$$

Here we have used the expansion $\ln u \approx -(1-u)$ for $u \approx 1$. Similarly, we can find the thermal exponent ν in the grand canonical ensemble:

$$R^2(K,\rho,w) = \frac{\sum_S [R(S)]^2 |S|^q K^{|S|} \rho^{\ell(S)} w^{V(S)}}{G_q(K,\rho,w)} \underset{K \to K_c(\rho,w)^-}{\sim} (K_c(\rho,w) - K)^{-2\nu}, \tag{2.10}$$

where $R(S)$ is the radius of gyration of surface configuration S.

3. Overview of Monte Carlo procedure. Sterling and Greensite [17] carried out the first major Monte Carlo simulation of the vesicle problem. In their algorithm one sweeps over the cubes of the lattice, checks if at least one face belongs to S. If so, a new trial surface S' is created by reversing the status of each face of this cube (i.e. the cube goes from inside to outside S or *vice versa*, with $V(S') - V(S) = \pm 1$). Before proceeding, they apply three tests (called R1, R2, and R3) to make sure that S' is connected, self-avoiding, and satisfies the topological constraints (if S already has these properties). Specifically, **R1** prevents disconnects in the form of isolated cubes in the interior or exterior by checking that at least one (but not all) plaquette of the trial cube belongs to S. The parenthetical condition also prevents S from disappearing entirely. The condition **R2** was mentioned in Section II in conjunction with closed surfaces: it requires that two and only two plaquettes meet at any link of the surface. For example, consider a chain of cubes with a right-angle bend or refer to Fig. 1. If the cube at the bend (labeled "R2" in Fig. 1) is removed, four plaquettes meet at the remaining link. This condition is computationally useful in keeping the checking local, but we shall see that its violation does not alter the universality class. Finally, **R3** forbids moves if two plaquettes on opposite sides of the trial cube do not belong to S, while the remaining four do, or *vice versa*. For example, in the chain of cubes mentioned above, one cannot remove a cube from the interior of a straight segment, such as the upper cube labeled "R3" in Fig. 1. Similarly, one cannot add the cube outlined with dashed edges in Fig. 1. This criterion prevents disconnects or changes of genus. (Finding which of these two possibilities occurs requires a global rather than a local check, as discussed after eqn. (7.1).) If all these conditions are met, S' is accepted if it passes the standard Metropolis test based on equilibrium distribution according to G.

Glaus [2,18] modified this algorithm in three ways. First, he fixed one plaquette of S by never visiting two adjacent cubes, one always inside S, the other always outside. Such a plaquette is labeled "G" in Fig. 1. This scheme added an extra factor of $|S| = n$ in the rate of transitions $S \to S'$. Second, rather than sweeping all cubes, he sampled—randomly—just those with faces on S. This idea not only improved efficiency but also added another extra factor of $|S|$. Third, he used a relatively large $20 \times 20 \times 20$ cube with pseudoperiodic boundary conditions, adjusting K to prevent excessive growth. The resulting probability of a transition in his

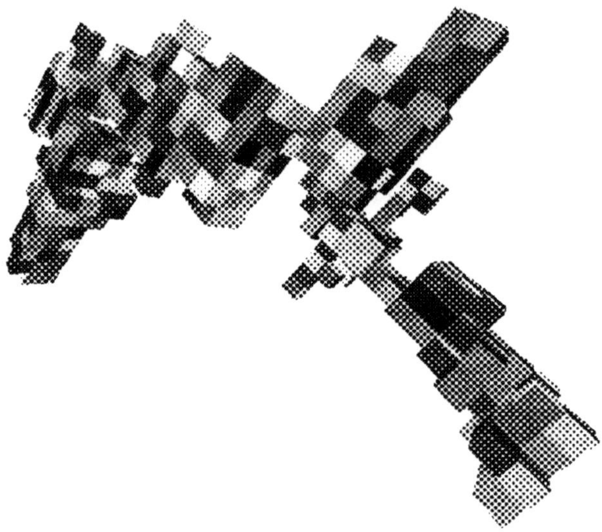

FIG. 2. *Sample of the configurations generated using the oct-tree data structure and imaged with Salem software on a Silicon Graphics Iris 4. Note the much greater size of the surface than that in Fig. 1, characteristic of earlier work.*

formulation is somewhat complicated, but can be found explicitly in refs. [2,18].

Orlandini *et al.* [6,19] developed a different set of modifications. Rather than sampling cubes, they chose at random a plaquette belonging to S; then picked at random the cube inside or outside with this face. To correct for the resulting bias toward cubes with multiple faces $\kappa_{SS'}$ on S, they rejected the move if $r > 1/\kappa_{SS'}$, where $r \in [0,1]$. Finally, rather than increasing the factors of n by "tricks," they simply included an extra factor n^q in the generating function. This formulation allows the transition probability to be written simply as

$$\omega(K, \rho = 1, w = 1; S \to S') = \begin{cases} A(S,S')|S'|^q & \textit{if } |S'| \leq |S| \\ A(S,S')|S'|^q K^{|S'|-|S|} & \textit{otherwise} \end{cases} \tag{3.1}$$

where $A(S,S') = (|S| + |S'|)^{-1}(|S|^q + |S'|^q)^{-1}$.

An important ingredient which leads to improved accuracy in the numerical study of SARS is the oct-tree data structure, described later in this volume by I. Beichl. Fig. 2 illustrates the large size of the clusters which are routinely generated in this approach.

4. Spherical vesicles without osmotic pressure or rigidity. The initial controversy in this area centered around the universality class of the model with neither osmotic pressure nor rigidity (i.e. $\rho = 1$ and $w = 1$), specifically on the values of the exponents θ and ν. For planar random surfaces Durhuus *et al.* [20] argued persuasively that the model collapses into non-interacting branched polymers in all dimensions, implying $\theta = 5/2$ and $\nu = 1/4$. These values were also obtained by Drouffe *et al.* [21] from a mean-field theory for SARS, expected to become exact for large dimensions. On the other hand, Baumann and Berg [22] calculated $\nu \simeq 1/4$ but $\theta \simeq 11/4$ for planar random surfaces without spikes in 4D, via accurate Monte Carlo methods, and thence concluded that the model does not collapse into non-interacting branched polymers. Using real-space renormalization, Maritan and Stella [23] calculated values for ν for a SARS model in $3 \leq d \leq 8$ that are considerably lower than those expected for branched polymers. Further support for this low ν came from a Flory argument [23]. As noted above, Sterling and Greensite [17] carried out the first Monte Carlo calculation of SARS. They found the value $\theta = 0.5 \pm 0.05$ for a fixed boundary in 3D. Subsequently, however, Glaus obtained from Monte Carlo estimates of the exponents close to $\nu = 1/2$ and $\theta = 3/2$, the values that Parisi and Sourlas [24] conjectured to be exact for branched polymers. Redner [25] found similar values by exactly enumerating [small] clusters for a model with a free boundary. Glaus and Einstein [18] reconfirmed the Monte Carlo result, quoting $\nu = 0.502 \pm 0.024$ and $\theta = 1.51 \pm 0.25$. Any glimmer of doubt was erased by improved simulations by O'Connell *et al.* [16], who found $\nu = 0.5029 \pm 0.004$ and $\theta = 1.500 \pm 0.026$. The physical idea behind this result is that spindly tree-like surfaces have sufficiently high entropy that they dominate behavior down to 3D, the lowest non-trivial dimension [2,26].

5. Vesicles with osmotic pressure or rigidity. If we allow for an osmotic pressure $\Delta p \equiv p_{in} - p_{out}$, we find a phase diagram as illustrated in Fig. 3. For negative Δp the vesicles remain finite for $K < K_c(\Delta p)$. As Δp becomes progressively more negative, the vesicles become more tubular and angular, with most of the branches approaching unit thickness. As one approaches $K_c(\Delta p)$ from below, they grow continuously larger, exhibiting the scaling properties of 3D branched polymers. For $K > K_c(\Delta p)$ we again have inflated, swollen vesicles that grow without bound. As one goes from negative to positive osmotic pressure, one encounters what appears to be a line of droplet-like essential singularities along the line segment $\Delta p = 0$ for $0 \leq K \leq K_c(0)$. Similar behavior was found for 2D vesicles by Fisher *et al.* [27]. As at the liquid-vapor transition, a vapor bubble, once formed, grows without bound since the bulk energy overwhelms the surface energy. This is a first-order transition: the vesicle size, i.e. the radius of gyration, does not diverge as $\Delta p \rightarrow 0^-$. However, it becomes infinitely large as soon as $\Delta p > 0$.

At $(\Delta p = 0, K_c(0))$ the curve of continuous branched polymer transitions joins the line of first-order transitions, so one expects tricritical

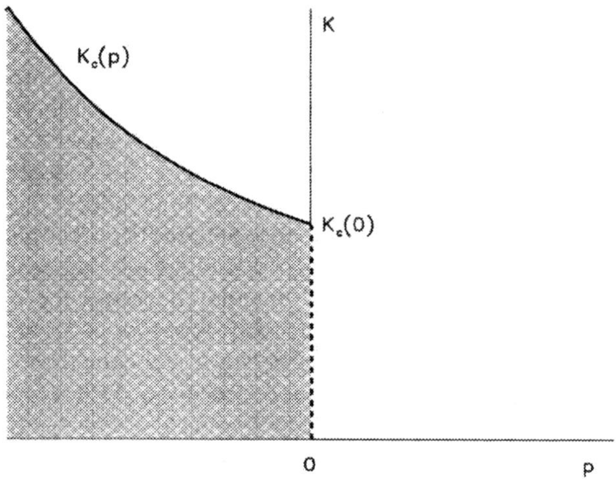

FIG. 3. *Phase diagram of plaquette fugacity K vs. pressure p, reproduced from Fig. 1 of ref. [28]. The generating function G is convergent only in the shaded region. Along the dashed vertical line there is a locus of droplet singularities. Further discussion is given in section V.*

behavior. However, we recall that this is just the transition point of vesicles with no osmotic pressure. Thus, the critical exponents should be those of branched polymers. By defining a new model in which R2 (defined in Section III) is violated (which we call *model 2*, in contrast to the original model, called now model 1), we can make connections with the behavior of interacting site lattice animals (LA). When we allow R2 moves, then the same edge can be shared by more than one pair (i.e. two pairs) of plaquettes. The edge then counts twice in $\ell(S)$. Denoting by m the number of cubes inside S, i.e. the volume $V(S)$, we can write the relation $|S| = 6m - 2l$, where l is the number of pairs of nearest-neighbor cubes (and so different from ℓ). Then we have

$$\tilde{G} \propto \sum_S K^{|S|} w^{V(S)} = \sum_S K^{6m-2l} w^m \equiv \sum_{LA} \lambda^m \mu^l . \tag{5.1}$$

In other words, model 2 can be mapped explicitly into interacting lattice animals (LA) in 3D. The center of each cube inside S is a site of the LA, with fugacity λ, and a Boltzmann factor μ is due to the interaction between each pair of nearest-neighbor sites of the LA. Under this mapping the point $(\Delta p = 0, K = K_c(0))$ can be interpreted as a Θ point of a particular type of site LA that lacks hollows in its configurations. (The Θ-point exponents are then indicated by ν_Θ and θ_Θ.) We find that at $\Delta p = 0$, $\nu = 0.506 \pm 0.005$ and $\nu = 0.48 \pm 0.01$ for models 1 and 2, respectively. From plots of $\langle |S| \rangle^{-1}$ vs. K, we find $\theta = 1.50 \pm 0.06$ and $\theta = 1.7 \pm 0.1$, respectively,

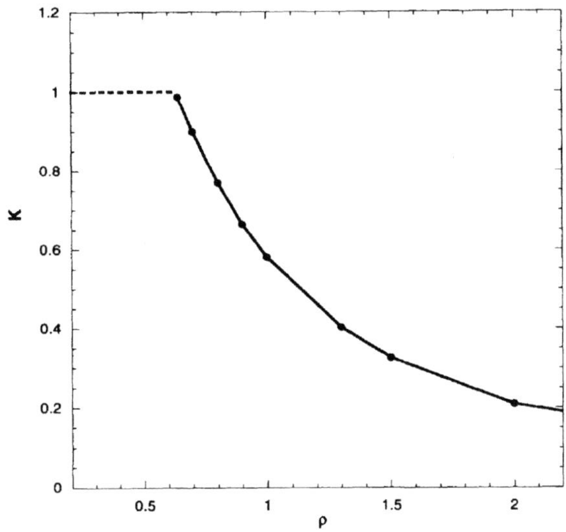

FIG. 4. *Phase diagram of plaquette fugacity* K *vs. bending rigidity* $\rho \equiv \exp(-\kappa/k_BT)$. *The dots, computed by Monte Carlo, are taken from Table 2 of ref. [29]. The solid curve indicates a continuous transition in the class of branched polymers. The dashed line indicates first-order transitions marked by droplet singularities. Above the phase boundary, vesicles diverge in size. We expect a tricritical point where the dashed and solid curves meet, near* $\rho = 0.63, K = 1$; *see text for discussion.*

still reasonably consistent with the picture of 3D branched polymers [24], as well as $\nu_\Theta = 1/2$ and $\theta_\Theta = 1.5$. Furthermore, the crossover exponent ϕ is $\phi_\Theta = 1$, as seen by the linear approach of $K_c(\Delta p)$ to $K_c(0)$.

6. Novel tricriticality of the crumpling point. In Fig. 4 we sketch the phase diagram when bending rigidity $\rho \equiv e^{-\kappa/k_BT}$ is included (but osmotic pressure is zero). Again, for $K < K_c(\rho)$ we find finite vesicles. For low to moderate rigidity (higher ρ), we find critical behavior characteristic of branched polymers as $K \to K_c(\rho)^-$.

In the perspective underlying Fig. 4, a first-order line changes into a second-order line at (ρ_c, K_c), leading to the expectation that this point is a tricritical point. In scans vs. ρ of critical properties, Orlandini *et al.* [29] find (cf. Table 2 therein) that, e.g., $\theta \approx 1.5$ down to $\rho = 0.7$, but at $\rho = 0.64$, θ changes to about 1.7. (These calculations were done to much higher accuracy than those leading to the report of $\theta \approx 1.7$ for model 2 in the previous section; the distinction from 1.5, relative to the small error bars, is significant here.) Similarly, $\nu \approx 0.5$ changes, albeit more modestly, to about 0.488. On the other hand, the surface-to-volume ratio and the length-to-surface ratio decrease gradually from 4 and 2 in the limit of large ρ to 2.14 and 0.98, respectively, at $\rho = 0.64$.

To gain further information about the critical properties at this point, we examined the divergences of several critical quantities as $K \to K_c^- = 1^-$ at our best estimate $\rho_c = 0.63$. The mean surface area area diverges with exponent $\theta = 1.78 \pm 0.03$, significantly larger than the branched-polymer value of 3/2. The mean-square radius of gyration diverges with exponent $\nu = 0.486 \pm 0.001$, slightly smaller than the value for branched polymers; in a canonical ensemble, this exponent is 0.482 ± 0.002. Since $\nu < 1/2$, the dominant configurations cannot be strictly wafer-like. Further evidence for this conclusion is that the ratio $\langle \ell(S) \rangle / \langle |S| \rangle$ is $\sim 0.930 \pm 0.002$ rather than zero, and the surface-to-volume ratio is $\sim 1.94 \pm 0.02$ rather than 2.

While the evidence in support of locating this tricriticality exactly at $(\rho_c, K_c(\rho_c) = 1)$ is rather compelling, and is corroborated by analogies with crumpled deflated ($\Delta p < 0$) vesicles, we note the very recent curious contradictory finding by E. J. Janse van Rensburg [30] that $K_c(\rho_c) < 0.9998^+$. The proof is summarized later in this volume. The resolution of this issue is still open. Given the minuscule difference of this bound from 1, it would be difficult if not impossible to resolve this controversy numerically. In any case, there must be a point of inflection in $K_c(\rho)$ very near our estimated ρ_c, and this may well be the precise location of the crumpling point.

7. Vesicles with handles and with laxer constraints. In all the previous discussion, the vesicles were restricted to zero genus, i.e. the topology of a sphere, with handles forbidden. Not only does this make the formal analysis simpler, we shall see shortly that the computational check for disconnects becomes much more taxing (i.e. it becomes non-local) if non-zero genus is allowed. The natural question is whether the universality class changes for unrestricted topology, i.e. if handles are permitted. Intriguing arguments in favor of such a change, based on renormalization-group ideas, were presented by Banavar *et al.* [31]. For example, the clusters were predicted to become more compact, causing the exponent ν to decrease and to approach 1/3 and the vesicles to resemble sponge-like structures.

In order to test these ideas, Stella *et al.* [28] carried out extensive simulations at fixed non-zero genus and at arbitrary genus. Associated with the genus is a genus fugacity L, so that the generating function becomes

$$G_q = \sum_S |S|^q K^{|S|} L^{H(S)}, \tag{7.1}$$

where $H(S)$ is the number of handles (the genus) of the surface S. At large L, vesicles with as many as 235 handles and thousands of plaquettes could be observed. When the genus is allowed to vary, the computations become particularly taxing since the check for disconnects (heretofore R3) becomes nonlocal: Imagine a portion of a vesicle consisting of a loop with a tail attached to the rest of the plaquette, both strands of unit cross-section.

If a cube along the handle is removed, the genus merely decreases by one and is allowed in this relaxed framework, but if a cube is removed from the tail, the loop is disconnected from the vesicle and the move is not allowed. There is no way to distinguish these two possibilities with a local check; a costly global graph check must be performed. F. Sullivan's article later in this volume elaborates on this issue.

In log-log plots of radius of gyration vs. area for fixed genera of 0, 1, 2, 3, and 4, the data turn out to fall on straight lines roughly parallel to each other: for the first three, the effective ν's are 0.532, 0.515, and 0.550, respectively. (No value was extracted for the other two due to the limited data range.) If the data from all 5 sets are combined, the effective exponent drops slightly to 0.481. However, when all genera are allowed (in a run of high handle fugacity $L = 16$), we find that at small area the data behave as before but at larger area it crosses over smoothly to an effective $\nu = 0.317$. Thus, the sponge-like compact behavior is observed only when arbitrary genus is allowed and very high genus promoted by the handle fugacity.

To resolve this apparent dilemma, we recall the scaling argument leading to the prediction of Banavar *et al.* [31]. They argue that

$$(7.2) \qquad \langle V\rangle_{\Delta p=0,K} \sim \left.\frac{\partial \ln G_q}{\partial \Delta p}\right|_{\Delta p=0} \underset{K\to K_c(0)^-}{\propto} (K_c(0)-K)^{-\phi},$$

with $\phi = 3\nu$. This suggests that critical SARS are dense, since they show that $\bar{d} = 3$ should then be the fractal dimension of the volume enclosed by the vesicle. In 2D, self-avoiding rings are in fact dense, and a similar renormalization-group argument leads correctly to this conclusion in that case [31]. From above, with $N \sim [K_c(0) - K]^{-1}$,

$$(7.3) \qquad \langle V\rangle_N/N \underset{N\to\infty}{\sim} N^{\phi-1} = N^{y\nu-1}$$

However, since $\langle V\rangle_N/N$ approaches a constant, $y\nu = 1$, implying that $\phi = 2\nu$, not 3ν. The resolution is that the renormalization argument of eqn. (7.2) fails when the proportionality factor implicit on the r. h. s. of eq. (13) vanishes, which happens in 3D but not 2D. This vanishing is beyond the predictive capabilities of the RG at the level of the arguments in ref. [31]. This property of 3D has been traced back to a concept of geometric redundancy of the osmotic pressure in the RG picture of 3D vesicles in the flaccid regime. Geometric redundancy implies that singular quantities lose their dependence on the scaling field proportional to Δp.

It is intriguing to explore analogies between surfaces with fixed genus at $\Delta p = 0$ and lattice animals (LA), particularly with regard to the influence of topology on the entropic critical exponent. The number of handles translates to the LA cyclomatic number c. In the LA case, for any fixed cyclomatic number, the critical fugacity K_c is the same as for $c = 0$. Likewise, for surfaces in model 1, neither ν (as just noted) nor K_c is very sensitive to H: $K_c^{H=0} = 0.577 \pm 0.004$, $K_c^{H=1} = 0.578 \pm 0.006$, and $K_c^{H=0} = 0.579 \pm 0.008$. For LA Soteros and Whittington [32] showed that $\theta_c = \theta - c$, whereas for SARS we find $\theta^{H=0} = 1.5$ (from earlier), $\theta^{H=1} = -0.5 \pm 0.5$, and $\theta^{H=2} = -2.3 \pm 1.0$. It is curious, to say the least, that these seem to differ by 2 rather than 1. However, the uncertainties in θ^H are very large.

8. Adsorption. In the adsorption problem one allows the surface to exist only in the positive half-space $z \geq 0$. Associated with adsorption there is a fugacity t, so that the generating function becomes

$$G_q(K,t) = \sum_S |S|^q K^{|S|} t^{N_a(S)}, \tag{8.1}$$

where N_a is the number of plaquettes lying on $z = 0$. This problem has been discussed by Orlandini *et al.* [33] and by Stella [4]. As $K \to K_c(t)^-$, we expect $\langle N_a \rangle \sim (K_c(t) - K)^{-\phi}$. There are several sorts of critical behavior (i.e. various possible ϕ's), analogous to the surface critical regimes of magnetic systems, reviewed e.g. by Binder [34]. One considers surfaces with a single plaquette fixed on the adsorbing plane. For small adsorption fugacity t, one has an ordinary regime analogous, as we have noted repeatedly, to 3D BP, not affected by the possibility of adsorption and so with $\phi = 0$. For higher t we find an adsorbed phase analogous to the surface transitions of magnetic systems. Here behavior is analogous to BP in *two* dimensions, with $\phi = 1$. Separating the two regimes at $t = t_1$ is a special transition with $\phi = \phi_s$, $0 < \phi_S < 1$. For very large fugacity, for $t > t_2 > t_1$, the vesicle is not only of unit height but seeks to minimize the perimeter of the adsorbed vesicle (i.e. the number of plaquettes with normals perpendicular to $\hat{z}$: the adsorbate changes from a 2D BP to a compact (monolayer-height) island, analogous to a 2D self-avoiding ring. At t_2 there appears to be a Θ collapse analogous to that of 2D interacting LA; the universality class of this transition still needs to be elucidated. It would be interesting to compare it to the other classes known for Θ regimes of branched polymers in 2D.

9. Summary and open questions. In conclusion, we now understand many of the properties of vesicles. Specifically,

1. For no osmotic pressure and no bending rigidity, the vesicles are flaccid and become critical for fugacity approaching the critical value $K_c(0)$ from below. The universality class is that of branched polymers, with $\nu = 1/2$ and $\theta = 3/2$.
2. The universality class remains that of branched polymers for negative osmotic pressure (deflated vesicles) and for moderate bending rigidity. Even for non-zero genus, this class describes behavior both when the number of handles is fixed at some finite value and when it is left free to fluctuate. On the other hand, for high handle fugacity a sponge-like morphology, with $\nu = 1/3$, is found.

3. For $K > K_c(\rho, w = 1)$ or for positive osmotic pressure in the absence of rigidity ($\rho = 1$, $w > 1$), the vesicles are described as expanded, swollen, and inflated in the non-equilibrium Monte Carlo runs. When bending rigidity is also included, the vesicles change from lamilar to cubic ("spherical" for the cubic lattice) as rigidity increases.
4. Several types of phase transitions are found in the various models:
   - Continuous [critical] (in the universality class of branched polymers).
   - First-order, with droplet singularities.
   - Tricritical, where they meet.
     - At zero osmotic pressure and bending rigidity: $K_c = 0.568$, analogous to the Θ point of lattice animals without holes, and apparently still in the universality class of branched polymers.
     - At $\rho_c = 0.635$ and $K_c = 1$, novel behavior: $\theta = 1.78 \pm 0.003$ and $\nu = 0.486 \pm 0.001$.

Advances in computing power and the formulation of clever algorithms now permit the numerical exploration of critical properties and the reliable determination of critical exponents for vesicles on lattices.

As examples of questions still remaining in this field, we mention the following:

- What is the precise law by which restricted genus determines entropic exponents in closed SARS? Is the relation between genus and θ consistent with that established between cyclomatic number and θ for lattice animals?
- The importance of knots is a major theme of this volume, but there is little, if any, evidence of their contribution to the configurations sampled in the reported simulations. As discussed later in this volume by F. Sullivan, there are lingering doubts about whether our algorithm can sample with adequate efficiency configurations that have high genus. Is it possible to construct a viable algorithm that samples more efficiently the phase space in which they lie? Would the results of the simulations, especially the critical properties, be altered significantly?

Acknowledgments. We gratefully acknowledge partial support from the U.S. Department of Defense and (TLE) from the U.S. National Science Foundation under MRG DMR 91-03031 and MRSEC DMR 96-32521.

REFERENCES

[1] J. Fröhlich, "The Statistical Mechanics of Surfaces," in: *Applications of Field Theory to Statistical Mechanics* (*Lectures Notes in Physics* **216**), ed. by L. Garrido (Springer, Berlin, 1985), 31–57.

[2] U. Glaus, "Monte Carlo Study of Self Avoiding Surfaces," J. Stat. Phys. **50**, 1141–1166 (1988).

[3] A.L. Stella, "Self Avoiding Surfaces and Vesicles on a Lattice," in: *Complex Systems: Fractals, Spin Glasses and Neural Networks*, ed. by G. Parisi, L. Pietronero, and M. Virasoro, Physica A **185**, 211–221 (1992).

[4] A.L. Stella, "Statistical Mechanics of Random Surfaces, Vesicles and Polymers," Turkish J. Phys. **18**, 244–260 (1994).

[5] A. Maritan, unpublished lecture notes.

[6] E. Orlandini, unpublished lecture notes.

[7] A. Maritan and A.L. Stella, "Some Exact Results for Self-Avoiding Random Surfaces," Nucl. Phys. **280**, 561–575 (1987).

[8] P.G. de Gennes, *Scaling Concepts in Polymer Physics* (Cornell University Press, Ithaca, 1979).

[9] P.M. Chaikin and T.C. Lubensky, *Principles of Condensed Matter Physics* (Cambridge University Press, Cambridge, 1995).

[10] S.G. Whittington, "Statistical Mechanics of Three Dimensional Vesicles," J. Math. Chem. **14**, 103–110 (1993).

[11] B. Durhuus, J. Fröhlich and T. Jonsson, "Self-Avoiding and Planar Random Surfaces on the Lattice," Nucl. Phys. B **225** **[FS9]**, 185–203 (1983).

[12] J.M. Hammersley, "The Number of Polygons on a Lattice," Proc. Camb. Phil. Soc. **57**, 516–523 (1961).

[13] A. Baumgärtner, "Inflated Vesicles: A Lattice Model," Physica A **190**, 63–74 (1992).

[14] A. Baumgärtner and A. Romero, "Microcanonical Simulation of Self-Avoiding Surfaces," Physica A **187**, 243–248 (1992).

[15] A. Baumgärtner, "Phase Transition of Semiflexible Lattice Vesicles," Physica A **192**, 550–561 (1993).

[16] J. O'Connell, F. Sullivan, D. Libes, E. Orlandini, M.C. Tesi, A.L. Stella and T.L. Einstein, "Self-Avoiding Random Surfaces: Monte Carlo Study Using Oct-Tree Data-Structures," J. Phys. A **24**, 4619–4635 (1991).

[17] T. Sterling and J. Greensite, "Entropy of Self-Avoiding Surfaces on the Lattice," Phys. Lett. B **121**, 345–348 (1983).

[18] U. Glaus and T.L. Einstein, "On the Universality Class of Planar Self-Avoiding Surfaces with Fixed Boundary," J. Phys. A **20**, L105–L111 (1987).

[19] E. Orlandini, Ph.D. thesis, U. of Bologna, 1993 (unpublished).

[20] B. Durhuus, J. Fröhlich and T. Jonsson, "Critical Behaviour in a Model of Planar Random Surfaces," Nucl. Phys. B **240** **[FS12]**, 453–480 (1984).

[21] J.M. Drouffe, G. Parisi and N. Sourlas, "Strong Coupling Phase in Lattice Gauge Theories at Large Dimension," Nucl. Phys. B **161**, 397–416 (1980).

[22] B. Baumann and B. Berg, "Non-Trivial Lattice Random Surfaces," Phys. Lett. **164**B, 131–135 (1985).

[23] A. Maritan and A.L. Stella, "Scaling Behavior of Self-Avoiding Random Surfaces," Phys. Rev. Lett. **53**, 123–126 (1984).

[24] G. Parisi and N. Sourlas, "Critical Behavior of Branched Polymers and the Lee-Yang Edge Singularity," Phys. Rev. Lett. **46**, 871–874 (1981).

[25] S. Redner, "Enumeration Study of Self-Avoiding Random Surfaces," J. Phys. A **18**, L723–L733 (1985); **19**, 3199 (E) (1986).

[26] M.E. Cates, "The Fractal Dimension and Connectivity of Random Surfaces," Phys. Lett. **161**B, 363–367 (1985);
H. Tasaki and T. Hara, "Collapse of Random Surfaces in the Connected Plaquettes Model," Phys. Lett. **112**A, 115–118 (1985).

[27] M.E. Fisher, A.J. Guttmann and S.G. Whittington, "Two-Dimensional Lattice Vesicles and Polygons," J. Phys. A **24**, 3095–3106 (1991).

[28] A.L. Stella, E. Orlandini, I. Beichl, F. Sullivan, M.C. Tesi and T.L. Einstein, "Self-Avoiding Surfaces, Topology, and Lattice Animals," Phys. Rev. Lett. **69**, 3650–3653 (1992).

[29] E. Orlandini, A.L. Stella, T.L. Einstein, M.C. Tesi, I. Beichl and F. Sullivan, "Bending-Rigidity-Driven Transitions and Crumpling-Point Scaling of Lattice Vesicles," Phys. Rev. E **53**, 5800–5807 (1996).

[30] E.J. Janse van Rensburg, "Crumpling Self-Avoiding Surfaces," J. Stat. Phys., Vol. 88 - No. 1/2, July 1997.

[31] J. Banavar, A. Maritan and A. Stella, "Geometry, Topology, and Universality of Random Surfaces," Science **252**, 825–827 (1991);
"Critical Behavior of Two Dimensional Vesicles in the Deflated Regime," Phys. Rev. A **43**, R5752–R5754 (1991) for the 2D case.

[32] C. Soteros and S. Whittington, "Critical Exponents for Lattice Animals with Fixed Cyclomatic Index," J. Phys. A: Math. Gen. **21**, 2187–2193 (1988).

[33] E. Orlandini, A.L. Stella, M.C. Tesi and F. Sullivan, "Vesicle Adsorption on a Plane: Scaling Regimes and Crossover Phenomena," Phys. Rev. E **48**, R4203–R4206 (1993).

[34] K. Binder, "Critical Behaviour at Surfaces," in: *Phase Transitions and Critical Phenomena*, vol. 8, ed. by C. Domb and J.L. Lebowitz (Academic, New York, 1983), chap. **1**, 1–144.

COMPUTATIONAL METHODS IN RANDOM SURFACE SIMULATION

ISABEL BEICHL* AND FRANCIS SULLIVAN†

1. Introduction. This paper explains how the simulation described in Einstein's paper in this volume was implemented in order to insure both correctness and efficiency. The physics is explained in Einstein's paper and the work on the entire project can be found in [1]. In this paper we explain the mathematics behind the computational datastructures.

2. Review of the process. Let's briefly review the process we are simulating. We start with a single cube and allow creation and deletion of new cubes producing structures such as those shown in Figures 1 and 2. A single step of the simulation is as follows: choose one of the faces of the cubes at random. In Figure 3 this is represented by the shaded square. At random we decide to *create* or *delete*, that is to create a new cube attached to the shaded cube or to delete the cube of which the shaded square is a face. The entire object must never become disconnected. To be connected to the object, a cube must meet the object along a face. Meeting along only an edge or only at a point is not considered a connection. So in Figure 3, the circle does not indicate a connection although it is legal configuration.

Throughout this paper, "plaquettes" and "faces" are used interchangeably. By SARS we mean the self avoiding random surface simulation, that is one where the genus never changes and SIRS is the self intersecting random surface where the genus is allowed to change. In neither case is a disconnect allowed.

The genus, in this case, is the number of "holes" or "handles". So a set of cubes in a ring would have genus 1 and a set of cubes that do not have any handles would be genus 0. For a more complete exposition of genus, see [4]. The example in Figure 3 has genus 0 because the connection at the circle is not considered a connection at all.

3. Problems to be solved. There are two problems we will address in this paper:

1) how do we represent the volume and surface efficiently as data in a computer and
2) how can we tell if a proposed move would make a disconnect of the surface?

4. Efficient volume representation: the oct-tree. To address the first problem, that of efficiently representing the volume, we use the oct-

* National Institute of Standards and Technology (NIST), 820/365, Gaithersburg, MD 20899.

† Center for Computing Sciences, 17100 Science Dr., Bowie, MD 20715.

tree [5],[6]. The oct-tree is a datastructure that stores the entire geometric object as one massive cube with many subdivisions which are also cubes. It is very efficient as far as space and consequently time go, because when a sub-cube is filled it is collapsed into a single cube with no sub-cubes. It is easiest to see with a picture. In Figure 4, when one cube grows to two cubes, a larger cube is made to encompass both. On the second line of Figure 4, when the larger cube, called an oct, is completely full, the smaller cubes (indeed octs) are no longer represented and only the large oct remains. The third line illustrates how when a structure grows outside of the current oct, one might have to create a new level of oct. In Figure 5 we show how this is represented computationally. There are three states that an oct can be in: completely full, completely empty and partially full/partially empty. In Figure 5 these are written as F, E, and P respectively. The octs are written as a tree with the largest oct at the root. If the oct represented at a node is F or E, then it will have no children in the oct-tree. If it is P, then it will be divided into 8 sub-octs, some of which will be full, empty or partially full/empty. The partially full/empty octs (P) will have children, none of the other oct nodes have children.

We used the oct-tree to represent the volume in both our SARS and SIRS calculations. However, in our implementation we made a special case of a cube that is on the boundary (the surface), in that we did not allow that oct to collapse into a larger unit. This is partly because we needed the information from boundary cubes to be available quickly, so we didn't want to have to un-collapse full octs in order to find them.

There are two advantages to using the oct-tree in this case. One is that we don't use as much memory by collapsing the full cubes. A full cube is stored as exactly one item. In addition, when we do need to search the entire structure we don't have to query every cube to find one particular. Instead we do a tree search, which is more efficient than a linear search.

5. How to tell if a move causes a disconnect: local versus global checking. Determining and maintaining connectivity is the major problem for the SIRS simulation. This is because, unlike the SARS simulation where connectivity can be checked locally, one MUST check globally for connectivity in the SIRS case. An illustration of this is in Figure 6. The SARS case is illustrated in the left side of Figure 6. If we had the situation of cubes 1, 2, 3, and 4 and wanted to delete cube 3, this would cause a disconnect and thus would not be allowed. However, we could discover that this move was illegal by noticing that cube 3 had neighbors only on the east and west sides and no others. We do not have to have a global search of the connectivity. Likewise for an insert. If we wanted to insert cube 12 into the bottom situation this would cause a change in genus (not allowed in the SARS case.) But once again one would only need to check locally. More details about the SARS simulation can be found in [3].

This local check is not possible in the SIRS case where the genus can indeed change. In the right side of Figure 6, let us suppose that we want to delete cube 3. We do NOT want the entire object to become disconnected. But it could be that the genus of the total object is 1 and cube 2 is connected to cube 4 by some sequence of cubes along the dotted line in the diagram, in which case removal of cube 3 would indeed be legal.

Our solution to this problem is to check for connectivity of the resulting complex globally for every putative move. We use putative moves because actually making the move and then undoing it is inefficient. More will follow on this later.

Another problem that faces the SIRS implementor is that BOTH delete and create can cause the boundary, (i.e., the surface) to disconnect. That the surface can disconnect on delete is obvious. But Figure 7 illustrates how this can be true for insert also. Suppose we want to insert a cube into a structure that is almost a 3×3×3 block of cubes but minus the very center cube and the one cube directly on top of the center cube. We want to create a cube at that very top center cube. The resulting volume is a 3×3×3 cube minus the center-most cube. The boundary of this object is disconnected namely two concentric boundaries of cubes, albeit ones of different sizes. Topologically, these are equivalent to two concentric 2-dimensional spheres — certainly not connected. Our solution to this problem is to check for connectivity before every possible delete AND every possible create.

6. How we assure connectivity. In this section we describe how we check for connectivity and explain why what we do really accomplishes the check. Our method is to maintain the dual graph of the boundary faces. This is done in the program as an adjacency list. For each face f, we associate a list of the faces that are connected to it. (Usually there are four such faces.) Figure 8 illustrates this. Here we show for face f, the black square with neighboring faces $g_1, \ldots, g_4$ the grey faces. There is for face f a list containing $g_1, \ldots, g_4$. Likewise face f appears in each of the lists for the g_i.

Let us consider the case of a create. (Delete is similar.) To create a cube, we look at those existing faces that the potential new cube would touched. In Figure 9, let us suppose that we wanted to create a new cube at the space with the dotted lines. In this figure we note that A and B, the shaded faces would be touched by the potential new cube. If we actually inserted the new cube, faces A and B would no longer exist. So we mark these faces as "out to lunch". In reality, we maintain an array indexed by the existing faces. A face is usually "in", but when we check for possible disconnect we take the potential faces to be deleted and change the array element to "out". This array is much easier to manipulate than an update on the entire oct-tree and its related datastructures. Then we use the connectivity list to see if the resulting structure is connected ignoring every face marked as "out". There is an exposition of connectivity checking in [2].

In the next section we will explain why it is correct to only consider faces that would be destroyed by a create or delete of a new cube and NOT include the new faces that would be created by a move.

Before that, there is an additional note that must be mentioned. Because we do not consider meeting at an edge or at a point to be a connection, we must also change the adjacency lists for the faces to reflect this, when considering potential creates and deletes. Figure 10a shows three cubes with the dotted cube being considered for deletion. Before being deleted, face $f1$ is a neighbor of face $f2$. But without the dotted cube this is no longer so. Thus, for checking connectivity, we might need to temporarily delete $f1$ from the adjacency list for $f2$ and likewise delete $f2$ from the adjacency list for $f1$. The same problem can occur for insertions. In Figure 10b, we are attempting to create the bold-faced cube. If we do this then face $f2$ becomes a neighbor of face $f1$ whereas previously they were not connected. So similar adjustments must be made to these adjacency lists.

7. Why does this work?. We simulate faces to be deleted but not faces to be inserted. We call the simulation of deleting a face a pseudo-delete, that is delete but only by using the "out-to-lunch" method described earlier. Pseudo-insert is similarly defined, but of course in the software we don't do pseudo-inserts. We now explain why this still gives the correct connectivity information.

There are two questions to be answered:

1) Is it possible that, after pseudo-deletes, we get a complex that is connected but WOULD be disconnected if we were to perform the pseudo-inserts also?
2) Is it possible that after pseudo-deletes we get a disconnected complex that WOULD be connected were we to do the pseudo-inserts?

The answer to question 1 is "no" and it is easy to see why. If we have something that is connected and add something that is connected to the original could it possibly disconnect? Clearly not.

Question 2 is more difficult. We consider two cases:

a) the case where the intersection of the potentially created or deleted cube intersects the rest of the complex in something that is simply connected (for example one square) or disjoint combinations of this case and
b) the intersection is not simply connected.

For case a, consider Figure 11. In this figure, we attempt to delete (creation is similar) the cube which intersects the remaining complex in exactly one face. Both examples on the top line are of this sort. If by pseudo-deleting the striped face we were to get a disconnected complex, could it possibly become connected by simulating the pseudo-insert of that face? No because if you look at the single face deleted, you could continuously deform what is left so that the hole is just the size of a point, and in a two-dimensional manifold, you will not make any difference in the connectivity by filling in the point or leaving it deleted. The question, would the resulting complex go from being disconnected to connected by filling in that point. No, it would make no difference.

This is the same situation with the examples from Figure 12. In each of these cases, the striped faces to be pseudo-deleted are deformable to a point. The same argument holds. These are the only ways to have case a.

The case in Figure 13 is actually, the same situation. Here we are attempting to delete the center cube. (This is an example of the disjoint combination of the simply connected case.) It intersects the remaining complex in two faces opposite each other. The question is, if this gives us something that is disconnected is it possible that it would really be connected if I pseudo-inserted the striped faces? In this case there are two squares that can be deformed to two separate points. And again the same argument as case a shows that this (doing pseudo-inserts of the points) would not change the connectivity in any way.

For case b there will seem to be two possibilities, but the first is really a special case of part a and the second does not occur. Case b is illustrated in the bottom two figures of Figure 13. Here we have a cube to be deleted that meets the remaining complex in a ring. The first example in the figure is really an example of case a because we assume in this model that there is no connection along an edge.

However, the second part of Figure 13 illustrates a cube being deleted whose intersection with the remaining complex is a ring. In this case if doing the pseudo deletes of the two top faces causes a disconnection of the surface then the surface must have been disconnected to begin with which we never allow. In other words this situation should never occur when doing a cube delete.

If this is a case of creating a cube, then we pseudo-delete the sides of the cube and not inserting the top and bottom. That this would cause a disconnect that would be connected if we put in the top and bottom squares, is impossible. The reason is again that by inserting or deleting a point in a two-manifold one cannot change the connectivity.

So, case b was actually vacuous. It is actually only another instance of the first case.

8. Euler characteristics. The Euler characteristic is defined for a complex to be the number of vertices - the number of edges + the number of faces. When you know the Euler characteristic, you know the genus although they are not equal. For our surfaces, the faces are the plaquettes. The edges are where one plaquette meets another plaquette. However we must be careful because when two cubes meet ONLY at an edge this is not considered a meeting, and so in that case we count these edges as two. Figure 14a is an example of an edge that needs to be counted twice. Similarly for vertices. The rest of Figure 14 lists some cases when a vertex is actually counted more than once. This is not difficult to do in practice because of the way in which the objects are represented as a datastructure.

Figure 15 illustrates some Euler characteristics. Note that 15a and 15b have the same Euler characteristic. This is because they both have the same genus, namely 0. Figure 15c has genus 1 and Euler characteristic 0. Genus 2 (two holes) will have Euler characteristic -2, and n holes will have Euler characteristic -n. More on the Euler characteristic can be found in [4].

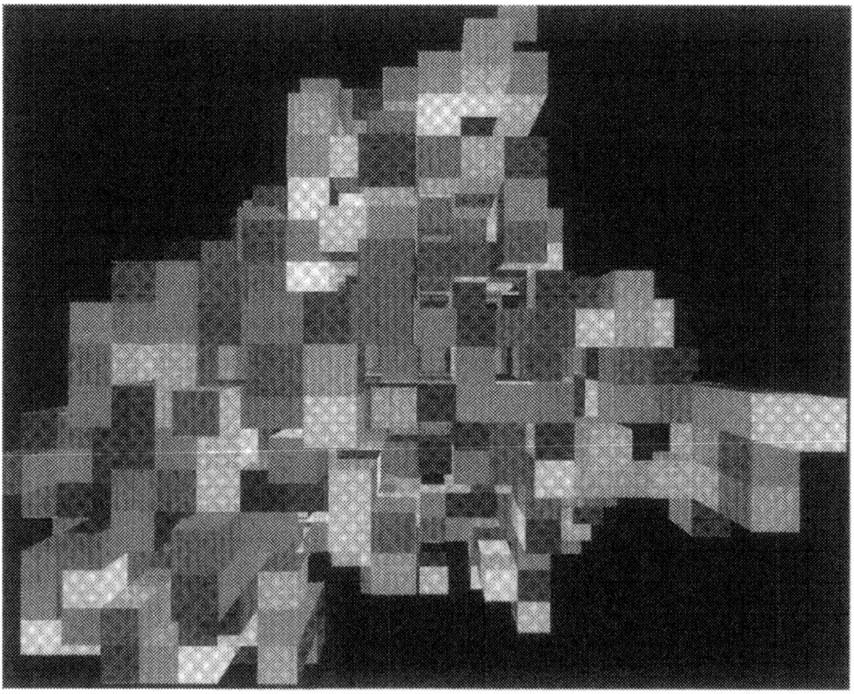

FIG. 1.

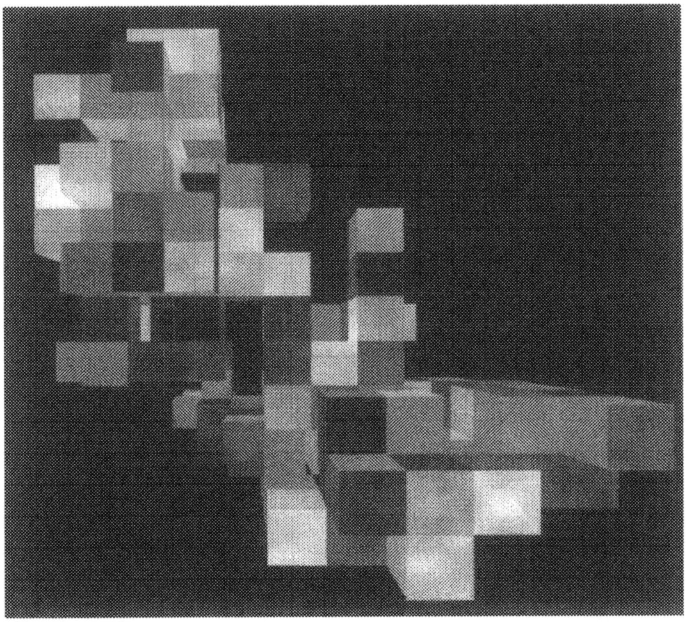

FIG. 2.

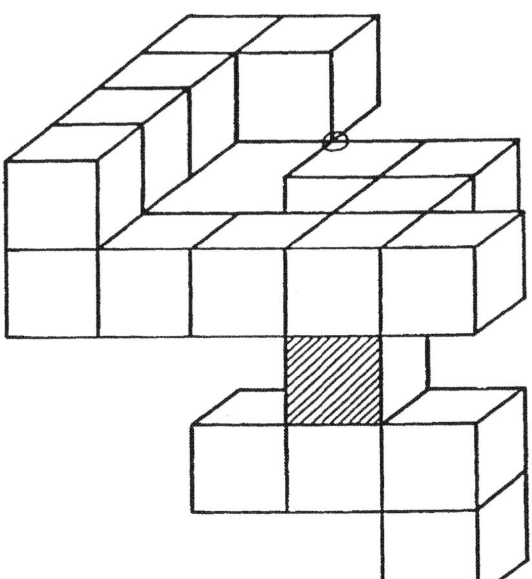

FIG. 3.

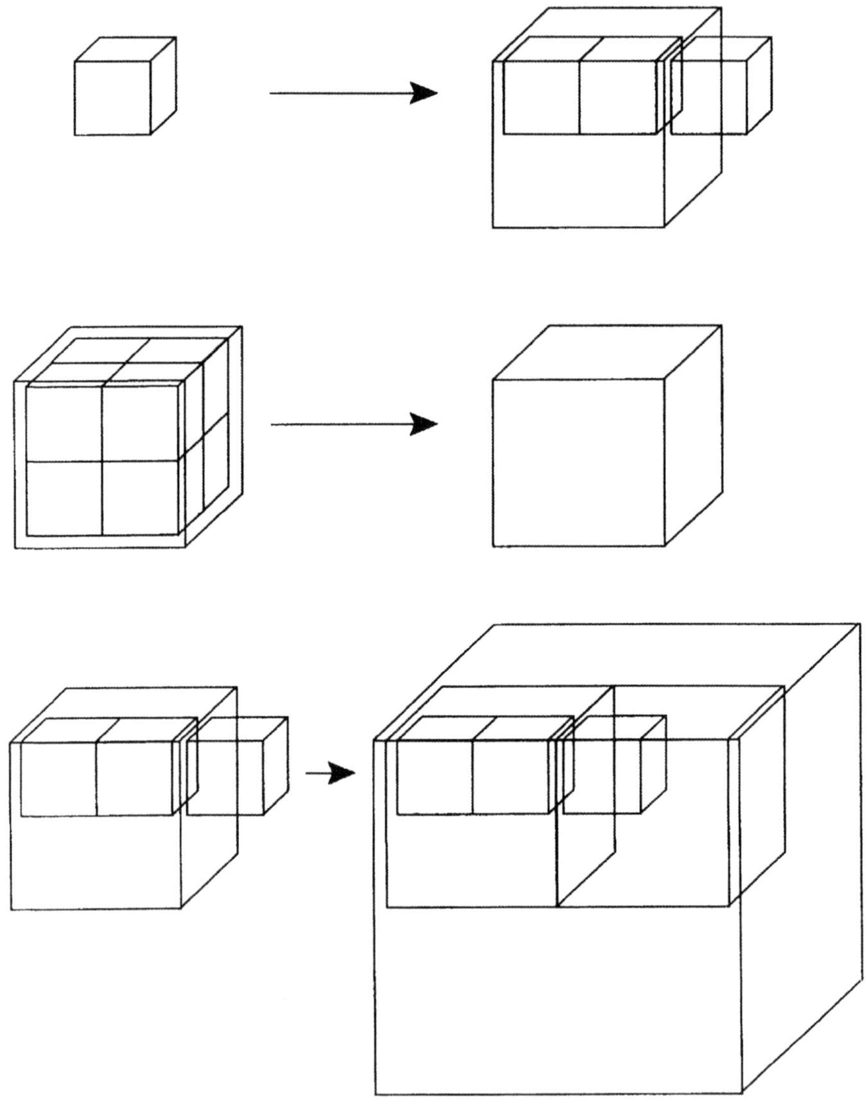

Fig. 4.

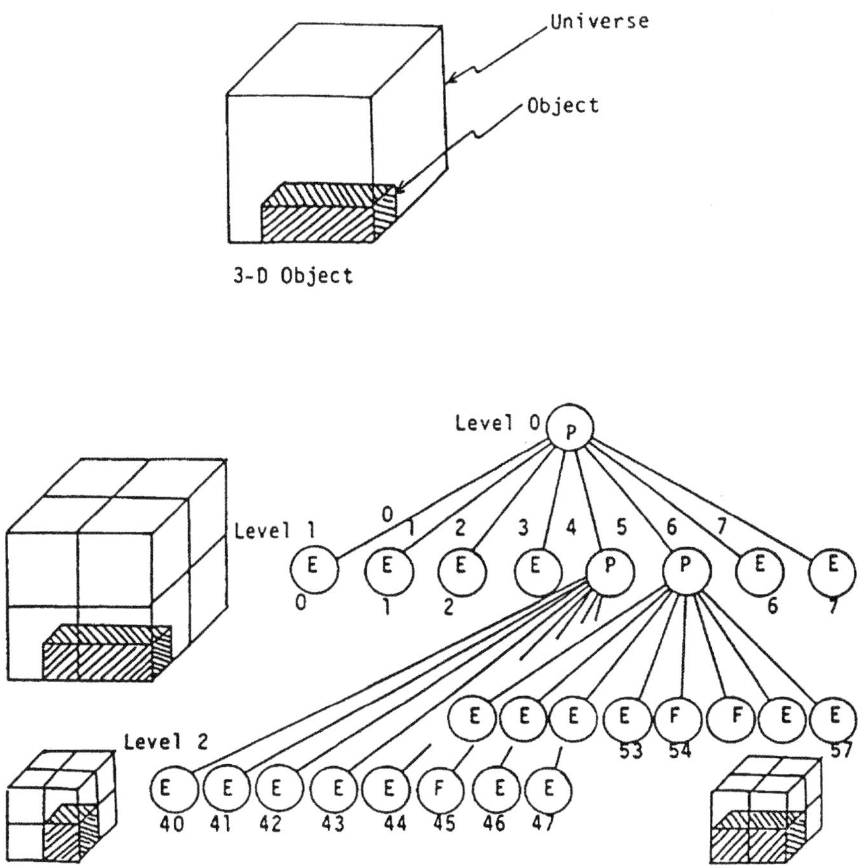

FIG. 5.

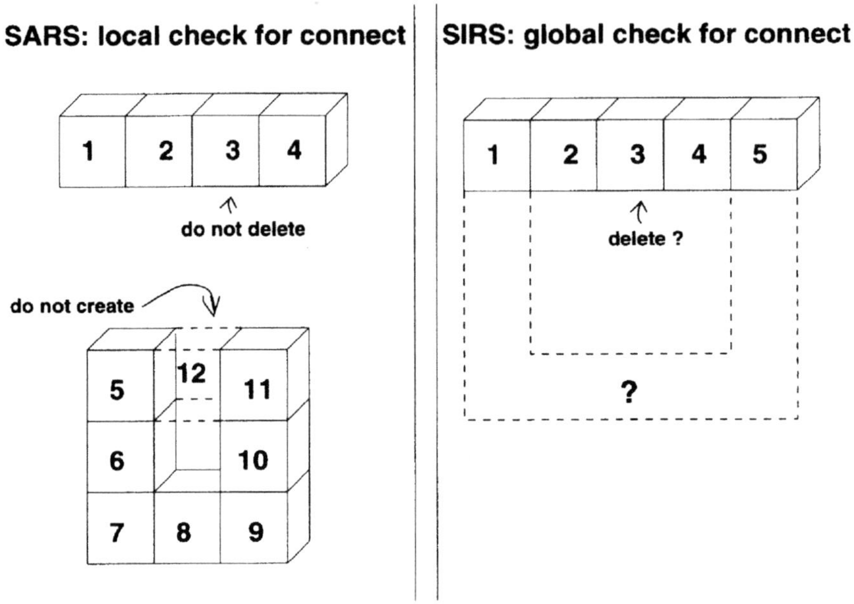
SARS: local check for connect
1
2
3
4
do not delete
do not create
5
12
11
6
10
7
8
9
SIRS: global check for connect
1
2
3
4
5
delete ?
?

FIG. 6.

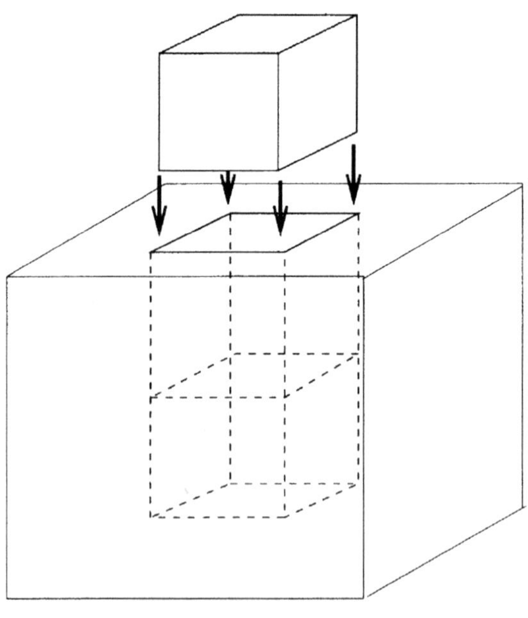

FIG. 7.

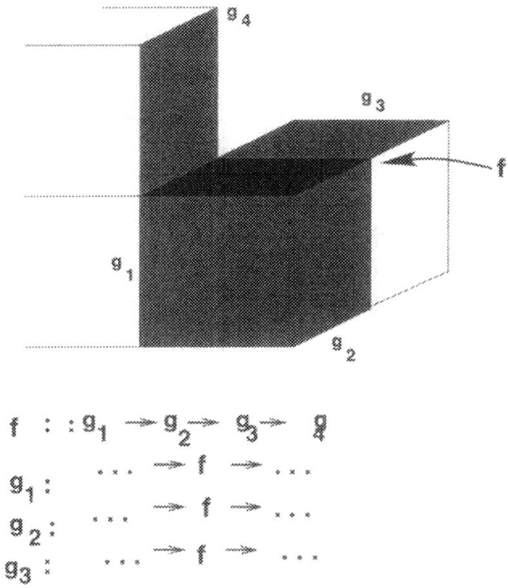
g4
g3
f
g1
g2
f : : g1 → g2 → g3 → g4
g1 : ... → f → ...
g2 : ... → f → ...
g3 : ... → f → ...

FIG. 8.

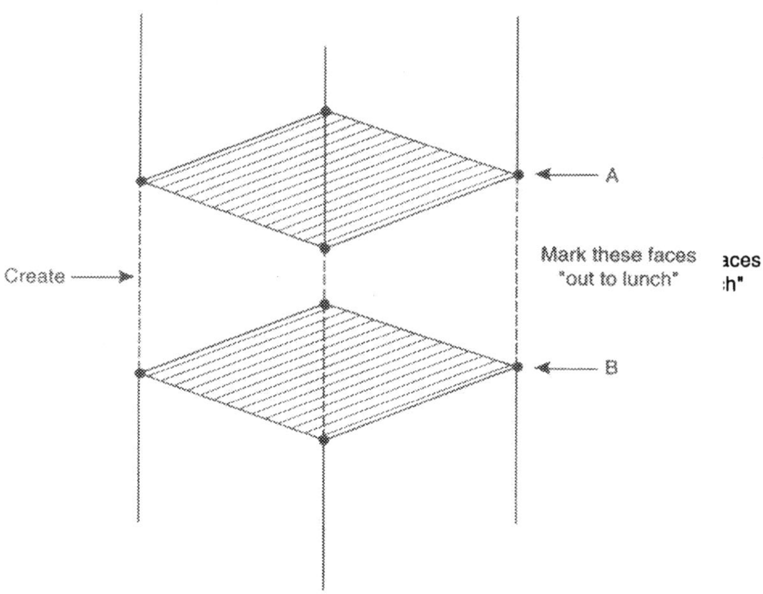
A
Create
Mark these faces
"out to lunch"
B

FIG. 9.

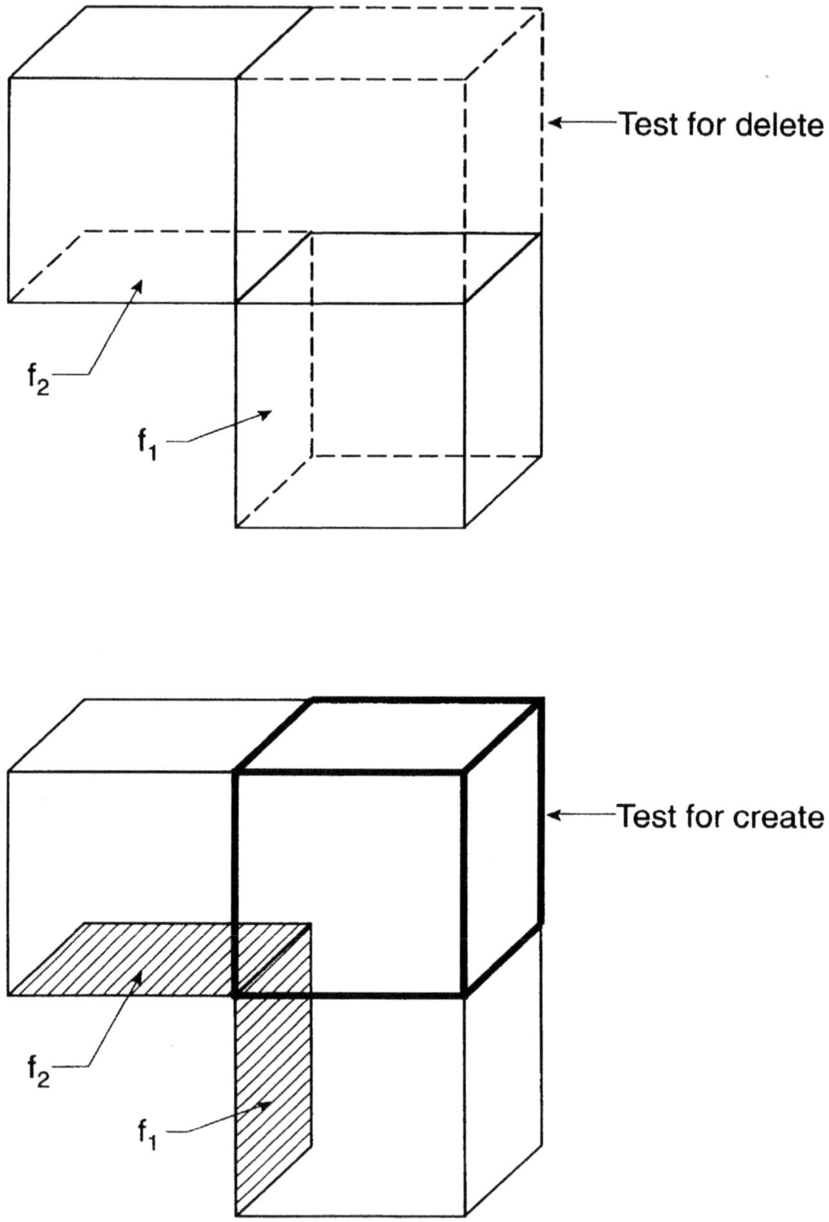

FIG. 10.

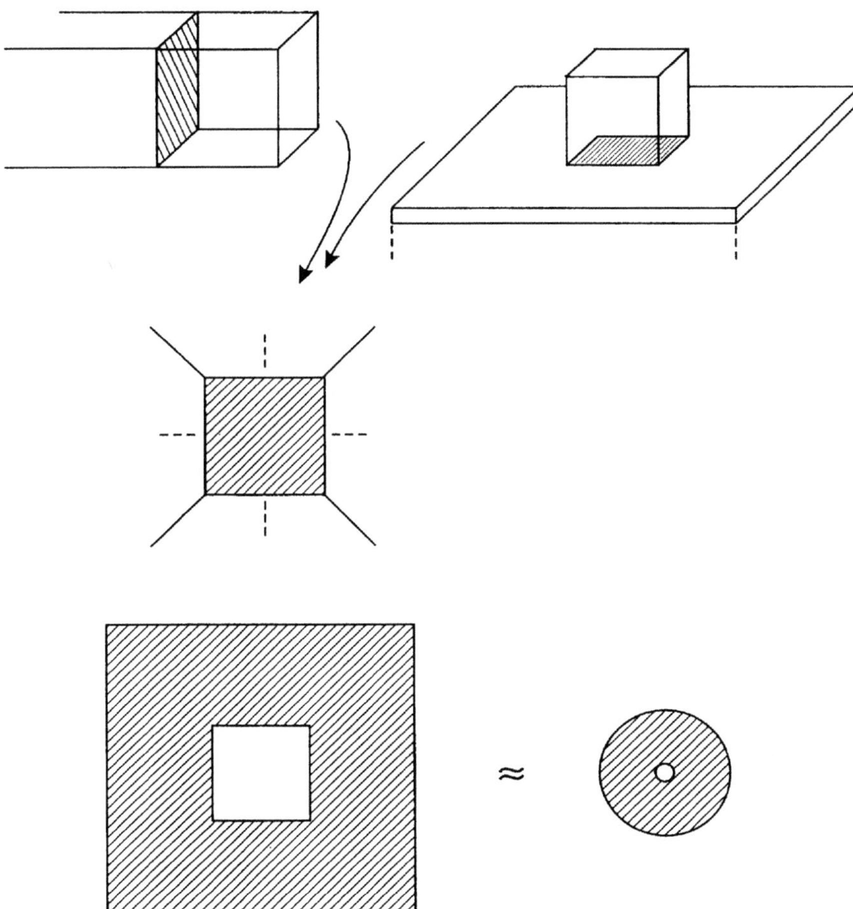

FIG. 11.

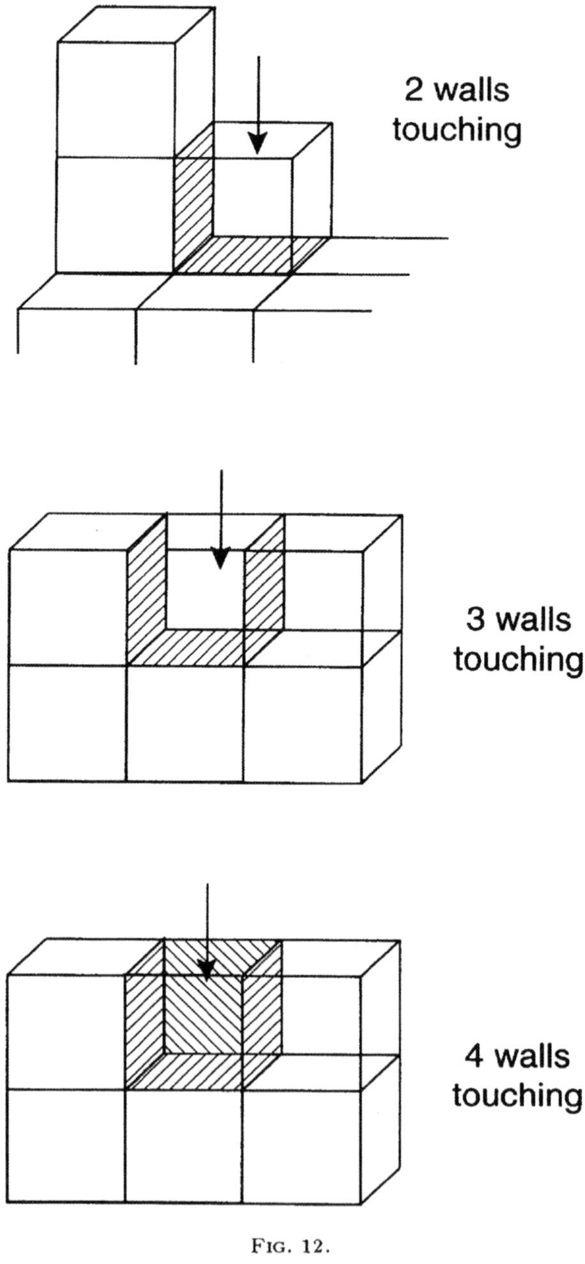

FIG. 12.

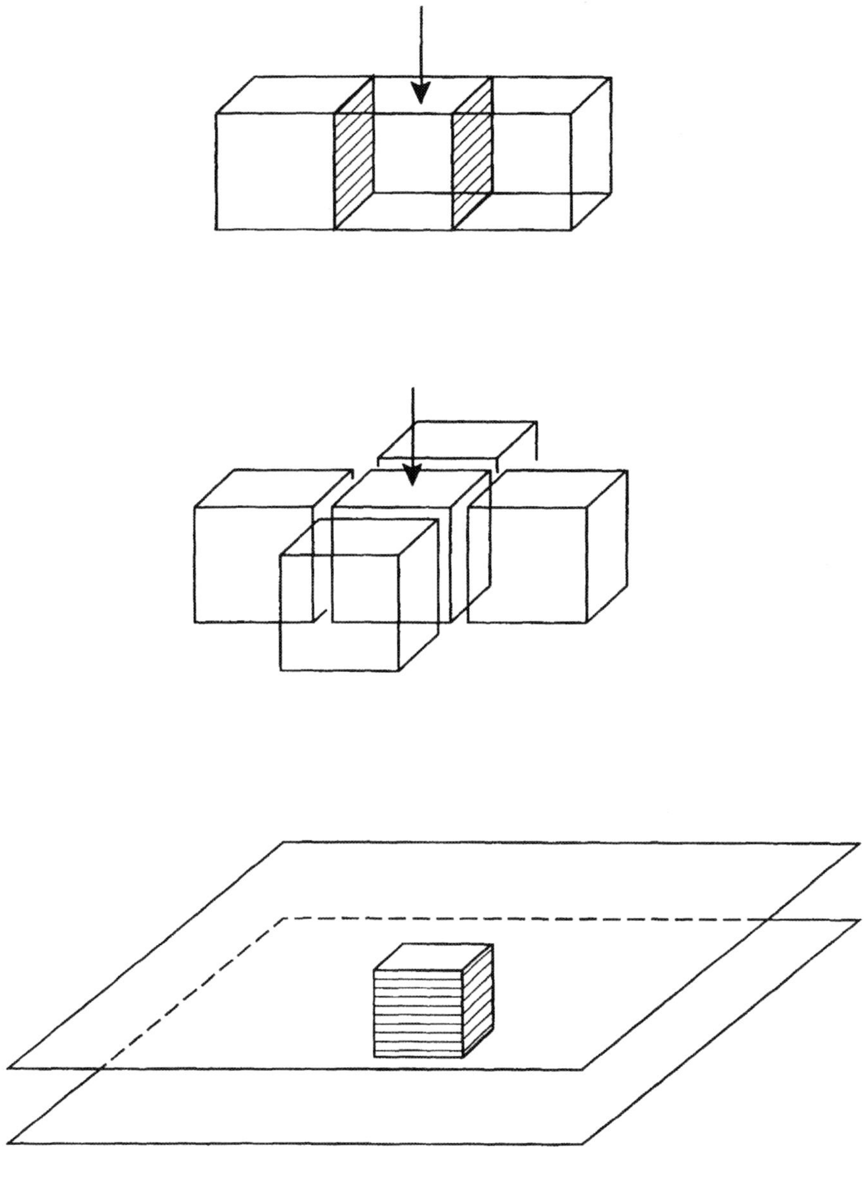

FIG. 13.

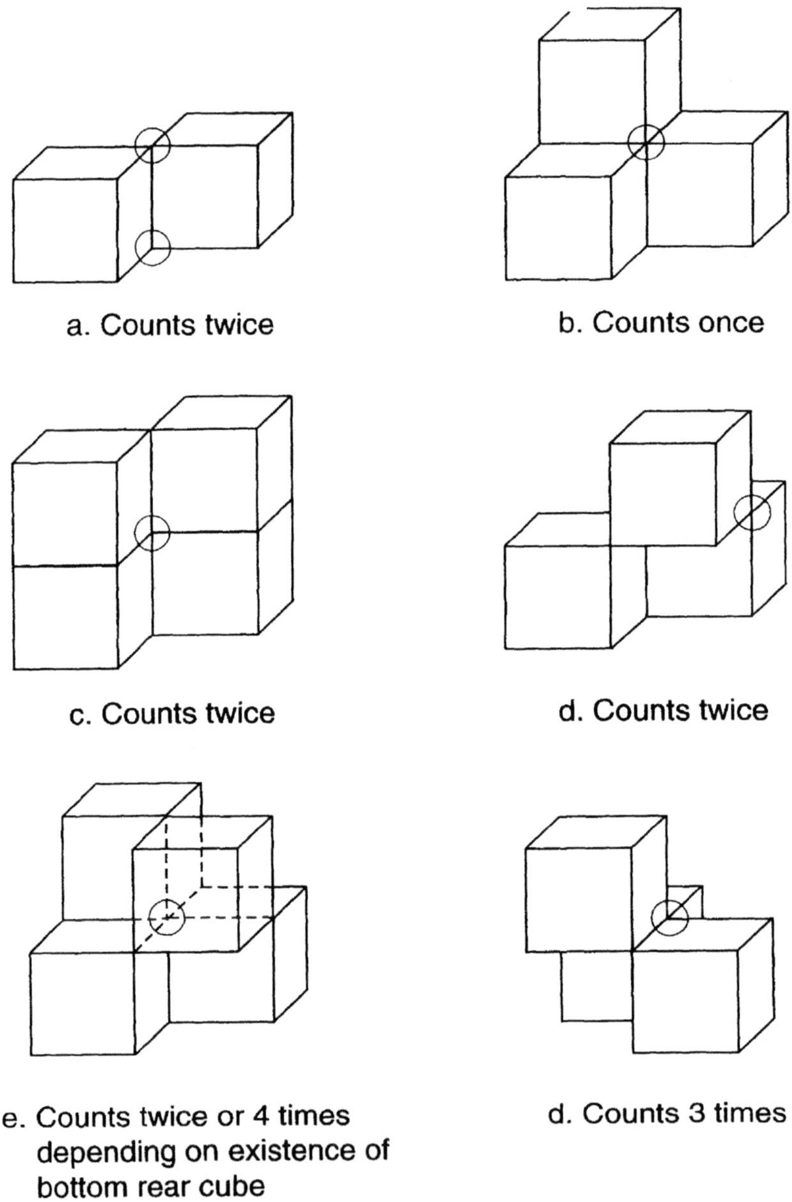
a. Counts twice
b. Counts once
c. Counts twice
d. Counts twice
e. Counts twice or 4 times depending on existence of bottom rear cube
d. Counts 3 times

FIG. 14.

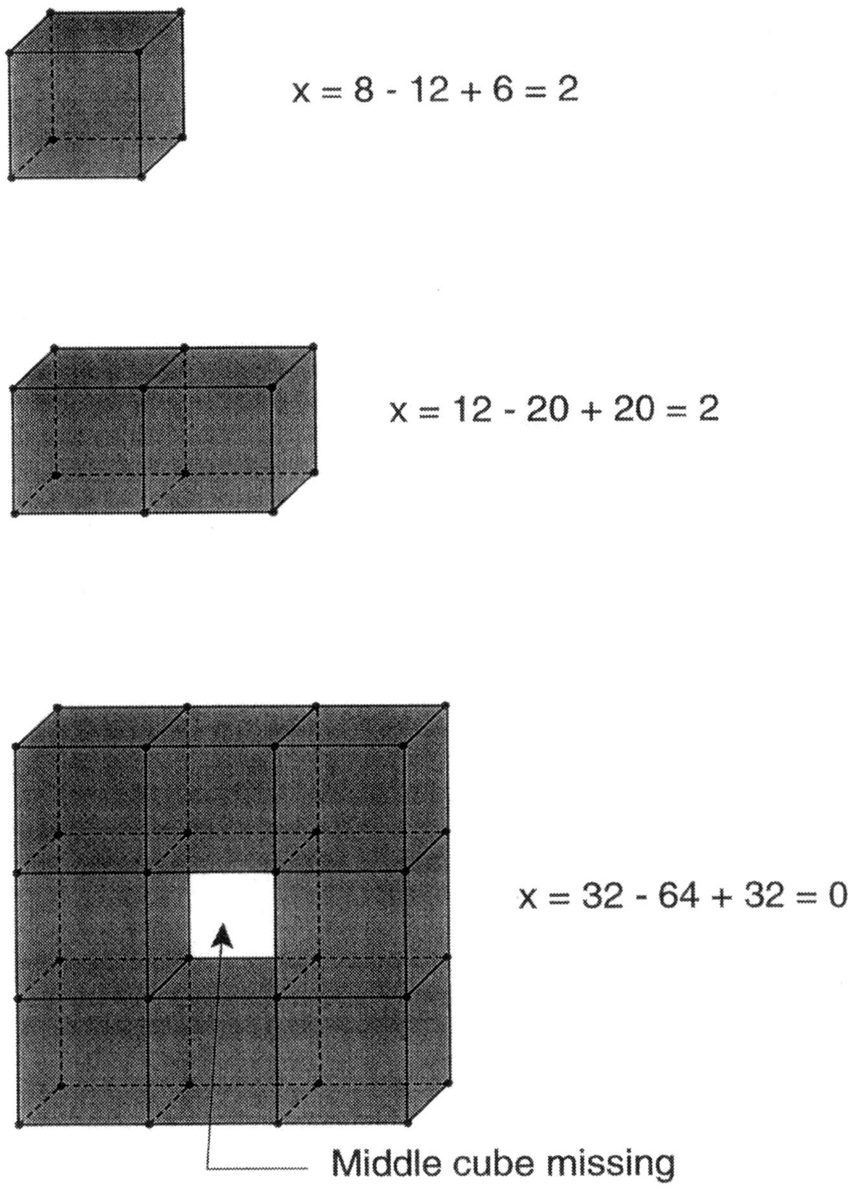

FIG. 15.

REFERENCES

[1] A.L. Stella, E. Orlandini, I. Beichl, F. Sullivan, M.C. Tesi and T.L. Einstein, *Self Avoiding Surfaces, Topology and Lattice Animals*, Phys. Rev. Lett. **69**, 3650 (1992).

[2] I. Beichl and F. Sullivan, *Making Connections*, IEEE Computational Sci. Engr. **3**, No. 3, 9 (1996).

[3] J. O'Connell, F. Sullivan, D. Libes, E. Orlandini, M.C. Tesi, A.L. Stella and T.L. Einstein, *Self Avoiding Random Surfaces: a Monte Carlo Study Using the Oct-tree Datastructure*, J. Phys. A **24**, 4619 (1991).

[4] J.G. Hocking and G.S. Young, *Topology* (1961).

[5] D. Libes, *Modeling Dynamic Surfaces with Oct-trees*, Computers and Graphics Magazine **15**, No. 3 (1991).

[6] H. Samet, *The Quadtree and Related Hierarchical Data Structures*, ACM Computing Surveys **16** (1984).

A MODEL OF LATTICE VESICLES

E.J. JANSE VAN RENSBURG*

Abstract. The phase diagram of a constrained lattice vesicle, called a c-surface, with a fugacity conjugate to the degree of bending and a fugacity conjugate to the volume of the vesicle, is derived and discussed. It is shown that the c-surface has at least three phases, which coexists at a triple point. These are an inflated phase, a crumpled phase, and a smooth phase.

1. Introduction. A *Lattice Vesicle* is a closed surface composed of plaquettes in the cubic lattice. Each plaquette is a unit square with vertices with integer coordinates, and edges of unit length between the vertices. The plaquettes are joined at edges, and at most two plaquettes can meet at an edge. The boundary of the surface is the set of edges incident with only one plaquette, and a surface is *closed* if it has no boundary. The surfaces are self-avoiding in the sense that each vertex has a unique set of integer coordinates. That is, if two plaquettes has a vertex with the same coordinates, then they must be adjacent through an edge incident with the vertex. See figure 1 for an example.

Lattice vesicles were introduced to describe the statistical mechanics of closed surfaces with the topology of a sphere. Lipid bilayer vesicles, other membranes in living systems, and microemulsions are examples of such surfaces [1,2]. Vesicle models for red blood cells, interfaces and other membranes have also been studied [3]. The phase diagram of vesicles is surprisingly rich. A vesicle can inflate or deflate under osmotic pressure [2], crumple or become smooth under surface tension or a curvature energy [4,5,6], or even assume other shapes under a variety of conditions [3].

The behaviour of lattice vesicles in the absence of a bending energy and with osmotic pressure is thought to be understood [7]. If the pressure difference between the interior and exterior of the vesicle is negative, then the vesicle deflates, and is said to be in the "flaccid" regime [2,8,9,10]. Numerical work suggests that the flaccid regime is characterized by branched polymer exponents [4]. If the pressure difference is positive, then the vesicle inflates; its volume scales as the $\frac{3}{2}$-power of its area [2,7]. Bending rigidity is believed to take vesicles through a "crumpling transition", which separates a phase of smooth vesicles from a crumpled phase [5,6].

c-Surfaces can be defined as follows: An edge is a *fold* if the plaquettes incident with it are perpendicular. The *skeleton* of a surface is its set of folds. I will call a surface with a *connected skeleton* a c-surface. (c-Surfaces are therefore a subset of surfaces, and they were studied in reference [6] with a fugacity conjugate to the degree of folding.) It is known that a c-surface undergoes a "crumpling transition" from a smooth phase (with

* Department of Mathematics and Statistics, York University, North York, Ontario M3J 1P3, Canada. Email: rensburg@mathstat.yorku.ca.

FIG. 1. *A c-surface and its skeleton. The skeleton is a lattice animal.*

low degree of folding) to a crumpled phase [6]. In this paper I discuss the phase diagram of c-surfaces in the presence of two fugacities: a fugacity conjugate to the number of folds (the "crumpling" or "folding" fugacity z), and a fugacity conjugate to the volume of the c-surface (the "pressure" or "volume" fugacity y). A c-surface is illustrated in figure 1.

2. The free energy of c-surfaces. The thermodynamic behaviour of a self-avoiding surface is described by its free energy. The partition function is obtained as follows: Consider two surfaces to be distinct if they cannot be made identical by a translation in the lattice. This defines equivalence classes of surfaces, and in each equivalence class the surfaces are identical up to a translation. Abuse the definition of a surface by calling the equivalence classes surfaces (or c-surfaces, whenever that applies). Let $s_n^c(m,l)$ be the number of c-surfaces with area n, volume m and l folds. Then the partition function is

$$Z_n(y,z) = \sum_{m,l} s_n^c(m,l) y^m z^l. \tag{2.1}$$

The free energy per plaquette is

$$F_n(y,z) = \frac{1}{n} \log Z_n(y,z). \tag{2.2}$$

The thermodynamic properties of a c-surface is also completely described by its thermodynamic potential, which is defined as $P(x,y,z) = -\log G(x,y,z)$, where $G(x,y,z)$ is the generating function of $Z_n(y,z)$:

$$G(x,y,z) = \sum_n Z_n(y,z) x^n. \tag{2.3}$$

In order to use these formal expressions, it is necessary that the existence of a thermodynamic limit as $n \to \infty$ is proven. Since $Z_n(y,z)$ is a polynomial, $F_n(y,z)$ is analytic, and there cannot be a phase transition at any finite value of n. Thus, any interesting phase behaviour will only be seen in the scaling limit, as $n \to \infty$.

The standard construction for proving the existence of the thermodynamic limit is to show that the partition function satisfies a generalized sub- or supermultiplicative relation [11,12]. Here it is achieved by concatenating two c-surfaces. The bottom and top vertices of a surface is found by a lexicographic ordering of the vertices. Similarly, the bottom and top plaquettes are found by a lexicographic ordering of the plaquettes by the coordinates of their midpoints. It can be shown that the top and bottom plaquettes are normal to the direction defined by the first coordinate [13]. Concatenation is performed by taking two surfaces and translating the second so that the first coordinate (of the midpoint) of its bottom plaquette is exactly one more than the first coordinate (of the midpoint) of the top plaquette of the first surface. The surfaces can be concatenated by deleting the top plaquette of the first, and the bottom plaquette of the second, and by adding four plaquettes to join them together. Details are given in references [7,2,6]. The result is

$$\sum_{\mu,\lambda} s^c_{n_1}(m-\mu, l-\lambda)s^c_{n_2}(\mu,\lambda) \leq \sum_{\mu=-2}^{2} s^c_{n_1+n_2+2}(m+1, l+2\mu). \tag{2.4}$$

Multiplying by $y^m z^l$ and summing over m and l gives the following supermultiplicative relation for $Z_n(y,z)$:

$$Z_{n_1}(y,z)Z_{n_2}(y,z) \leq y^{-1}\Big[\sum_{\mu=-2}^{2} z^{2\mu}\Big] Z_{n_1+n_2+2}(y,z). \tag{2.5}$$

This result is enough to prove the following:

THEOREM 2.1. *There exists a limiting free energy in a model of c-surfaces with partition function (2.1). That is*

$$\lim_{n\to\infty} F_n(y,z) = \mathcal{F}(y,z)$$

exists for every y and z (but may be infinite).

Proof. By equation (2.5) it is apparent that $Z_n(y,z)/(y^{-1}\sum_{\mu=-2}^{2} z^{2\mu})$ is a super multiplicative function of the kind considered in reference [12]. Thus the limit exists, and if it is finite, then

$$F_n(y,z) \leq \Big(y^{-1}\sum_{\mu=-2}^{2} z^{2\mu}\Big)[\mathcal{F}(y,z)]^{n+2}. \quad \square$$

Remark. Since $Z_n(y,z)$ is a polynomial in each of y and z, it follows from the Cauchy-Schwartz inequality that $\log Z_n(y,z)$ is a convex function of both $\log y$ and $\log z$. In these circumstances $\log Z_n(y,z)$ is a *doubly convex* function of $\log y$ and $\log z$, and it is continuous for $y \in (0,\infty)$ and $z \in (0,\infty)$ (since it is finite). The free energy $\mathcal{F}(y,z)$ is the limit of a sequence of doubly convex functions, and is also continuous for $y \in (0,\infty)$

and $z \in (0, \infty)$ *provided that it is finite.* I will later indicate when the free energy is finite. It will also turn out that $\mathcal{F}(y, z)$ is a non-analytic function, and its non-analyticities will correspond to phase transitions in the c-surface. The radius of convergence of the generating function in equation (2.3) is obtained by the root-test:

$$x_c(y, z) = \lim_{n \to \infty} [Z_n(y, z)]^{-1/n} = e^{-\mathcal{F}(y,z)}. \tag{2.6}$$

In other words, the limiting free energy determines the radius of convergence of the thermodynamic potential $P(x, y, z)$ of the model. Moreover, the nature of the singularities in the thermodynamic potential, and the free energy, will determine the order and properties of the transitions.

3. The phase diagram of a c-surface. In this section the phase diagram of a c-surfaces with a volume and bending fugacity is discussed. Let $y = 1$ in (2.1) and (2.2). Then (2.1) is the partition function of a c-surface with a bending fugacity z. This model was considered in reference [6], and the following is known:

THEOREM 3.1. *There exists a critical value z_c such that $x_c(1, z) = 1$ if $z \leq z_c$ and $x_c(1, z) < 1$ otherwise. In other words, $\mathcal{F}(1, z) = 0$ if $z \leq z_c$ and $\infty > \mathcal{F}(1, z) > 0$ otherwise.*

By theorem 3.1 there is a non-analyticity in $\mathcal{F}(y, z)$ if $y = 1$ and $z = z_c$. The *degree of crumpling* is defined as

$$\frac{\partial \mathcal{F}(1, z)}{\partial z} \begin{cases} = 0, & \text{if } z < z_c \, , \\ > 0, & \text{if } z > z_c \, . \end{cases} \tag{3.1}$$

In other words, the c-surface is "smooth" (with a zero density of folds) if $z < z_c$ and $y = 1$. For larger values of z the c-surface has a positive density of folds. It is believed that the c-surface collapses to a branched polymer if $z > z_c$. On the other hand, if $z < z_c$, then a phase of cubical or disk-like c-surfaces may be found [5,6].

Suppose now that $y > 1$ instead. A q-cube is a c-surface which is a geometric cube of sidelength q. It has area $6q^2 = n$, volume $q^3 = (n/6)^{3/2}$ and number of folds $12q = 12(n/6)^{1/2}$. If the limit in theorem 2.1 is taken through q-cubes, then

$$\mathcal{F}(y, z) \geq \lim_{q \to \infty} \frac{1}{6q^2} (q^3 \log y + 12q \log z) = \infty, \tag{3.2}$$

since $y > 1$. Thus, the limiting free energy is divergent if $y > 1$, regardless of the value of z. Since $\mathcal{F}(1, z) < \infty$ for any $z < \infty$, there is a line of phase transitions along the line $y = 1$. If $z < z_c$, then the transition is from a phase of a smooth c-surface to an inflated c-surface, and if $z > z_c$, then the transition is from a phase of branched polymers to an inflated c-surface.

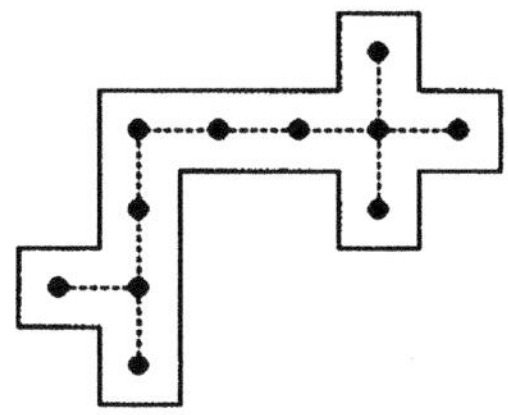

FIG. 2. *A c-surface with minimal volume is dual to a site tree.*

To see that there is indeed a phase of an inflated c-surface if $y > 1$, argue as follows: First note that (the limit is taken through a sequence of q-cubes)

$$\lim_{q\to\infty} \frac{1}{q^3} \log Z_{6q^2}(y,z) \geq \lim_{q\to\infty} \frac{1}{q^3}(q^3 \log y + 12q \log z) = \log y, \tag{3.3}$$

On the other hand, note that $Z_n(y,z) \leq s_n^c y^{(n/6)^{3/2}}[\max_l z^l]$, where we have replaced m and l by values which maximizes the powers, and where the sums over m and l has been executed. Since s_n^c grows at most exponentially with n [6], and $0 \leq l \leq 2n$,

$$\begin{aligned}
&\lim_{n\to\infty} \frac{1}{(n/6)^{3/2}} \log Z_n(y,z) \\
&\leq \lim_{n\to\infty} \frac{1}{(n/6)^{3/2}}\Big(\log\big(s_n^c y^{(n/6)^{3/2}}[\max_l z^l]\big)\Big) = \log y.
\end{aligned} \tag{3.4}$$

By (3.3) and (3.4), the partition function is dominated by surfaces of volume $(n/6)^{3/2}$; that is, inflated c-surfaces. Note that the derivatives

$$\left.\frac{\partial^N \mathcal{F}(y,z)}{\partial y^N}\right|_{y=1} < \infty,$$

in other words, there is an essential singularity in $G(x,y,z)$ along the line $y = 1$. The transition to an inflated c-surface is therefore a first order transition, and one can argue along the lines of reference [7] (see also [14]) that these are "droplet singularities" in analogy with condensation of a fluid along the line $y = 1$. These droplets coalesce as y increases from $y = 1$ into the inflated phase.

The situation is more complicated when $y < 1$. One notes that $\mathcal{F}(y,z) \leq \mathcal{F}(1,z) < \infty$ in this case (by theorem 2.1). In order to derive bounds on $\mathcal{F}(y,z)$, I define *deflated c-surfaces* to be that set of c-surfaces with volume $m = \frac{n-2}{4}$. Let the number of such c-surfaces be $s_n^d(l) = s_n^c(\frac{n-2}{4}, l)$. (Note that $\min\{l\} = n+6$, in which case the c-surface is a long thin (capped) cylinder.) I illustrate a 2 dimensional cut through such a surface with volume $\frac{n-2}{4}$ in figure 2. These c-surfaces are dual to a subset of site-trees, and by concatenation, and the techniques of theorem 2.1, it is seen that the limiting free energy $\mathcal{F}^d(z)$ of deflated c-surfaces exists and is defined by

$$\mathcal{F}^d(z) = \lim_{n\to\infty} \frac{1}{n} \log \sum_l s_n^d(l) z^l. \tag{3.5}$$

Moreover, using theorem 10 from reference [6], I conclude that there exists critical z_d such that

$$\mathcal{F}^d(z) \quad \begin{cases} > 0, & \text{if } z > z_d\ , \\ = 0, & \text{otherwise}\ . \end{cases} \tag{3.6}$$

Since, $s^d(l) \leq s^c(l)$, $z_d \geq z_c$. Observe that

$$\mathcal{F}^d(z) \geq \lim_{n\to\infty} \frac{1}{n} \log(z^{n+6} s_n^d(n+6)) \geq \log z$$

if $z > 1$. Hence z_d is finite. The fact that $\mathcal{F}^d(z) = 0$ if $z < z_d$ can also be derived as follows: Note that $s_n^d(l) \leq 4^l a_l$, where a_l is the number of lattice animals with l edges. This inequality is obtained by removing all plaquettes from surfaces counted by $s_n^d(l)$, leaving behind only the skeleton, which is a lattice animal. The factor 4^l appears because the number of ways of putting back a plaquette on an edge in the lattice animal is 4. Thus $\sum_l s_n^d(l) z^l \leq \sum_l 4^l a_l z^l < \infty$ if $z < (4\lambda)^{-1}$, where λ is the growth constant of lattice animals. For the generating function one finds $G^d(x,z) = \sum_n (\sum_l s_n^d(l) z^l) x^n \leq \sum_n (\sum_l^\infty s_n^d(l) z^l) x^n < \infty$ if $z < (4\lambda)^{-1}$ and $x \leq 1$. Thus, if $z < (4\lambda)^{-1}$ then the radius of convergence of the generating function is $x_c^d(z) = 1$, and $\mathcal{F}^d(z) = 0$, as claimed in equation (3.6).

A lower bound on the free energy $\mathcal{F}(y,z)$ can be derived as follows if $y < 1$:

$$\begin{aligned} Z_n(y,z) &= \sum_{m,l} s_n^c(m,l) y^m z^l, \\ &\geq \sum_l s_n^d(l) y^{(n-2)/4} z^l, \quad \text{taking } m = (n-2)/4, \\ &= y^{(n-2)/4} \sum_l s_n^d(l) z^l. \end{aligned} \tag{3.7}$$

Using (3.5) it is then found that

$$\mathcal{F}(y,z) \geq \frac{1}{4} \log y + \mathcal{F}^d(z), \quad \text{if } y < 1. \tag{3.8}$$

On the other hand, an upper bound is found more easily:

$$
\begin{aligned}
Z_n(y,z) &= \sum_{m,l} s_n^c(m,l) y^m z^l, \\
&\leq \sum_{m,l} s_n^c(m,l) y^{(n-2)/4} z^l, \quad \text{since } \min(m) = (n-2)/4, \\
&= y^{(n-2)/4} \sum_{l} s_n^c(l) z^l.
\end{aligned} \tag{3.9}
$$

Taking logarithms, dividing by n and letting $n \to \infty$ gives

$$
\mathcal{F}(y,z) \leq \frac{1}{4} \log y + \mathcal{F}(1,z). \tag{3.10}
$$

By theorem 3.1, (3.6), (3.8) and (3.10) one finds that

$$
\mathcal{F}(y,z) \begin{cases} = \dfrac{1}{4} \log y, & \text{if} \quad z < \min\{z_c, z_d\}, \\ \geq \dfrac{1}{4} \log y + \log z, & \text{if} \quad z > 1. \end{cases} \tag{3.11}
$$

Thus, for every $y < 1$ there is a non-analyticity in $\mathcal{F}(y,z)$ corresponding to a critical value in z (and thus to a "crumpling transition", where the c-surface becomes smooth). These results can be taken together by drawing the phase diagram of a c-surface, which is figure 3.

4. Discussion. I have shown that the phase diagram of a c-surface has at least three phases. If the osmotic pressure is positive, then the c-surface inflates to a phase with large volume, the presence or absence of a bending energy does not affect this. This was also shown by Whittington [2]. If the pressure fugacity is put equal to 1 ($y = 1$), then there are two phases. The first is a smooth phase of c-surfaces (believed to be cubical or disk-like [5,6]) at low values of the bending fugacity, and a "flaccid phase" or branch polymer phase at high values of the bending fugacity.

If the pressure fugacity is smaller than 1 ($y < 1$, or at negative osmotic pressure), then there are at least two phases. The first is a smooth phase, which is suggested to be a phase of rod-like vesicles by equations (3.8) and (3.10). It is not clear that this phase is different from the cubical or disk-like vesicles encountered in the smooth phase when $y = 1$, but the deflation of cubical or disk-like vesicles to rod-like vesicles seems to be likely to be a transition. The second phase is a phase of deflated crumpled vesicles, with branch polymer characteristics. If $y = 1$, then the best lower bound for large z is obtained by inflating the vesicles. This is in contrast to the deflated branched vesicles obtained here for $y < 1$.

In summary, there seems to be three broadly defined cases: The first is at $y > 1$, when a phase a inflated vesicles is encountered. The second is at $y = 1$, where two phases of *inflated* smooth and *inflated* crumpled vesicles

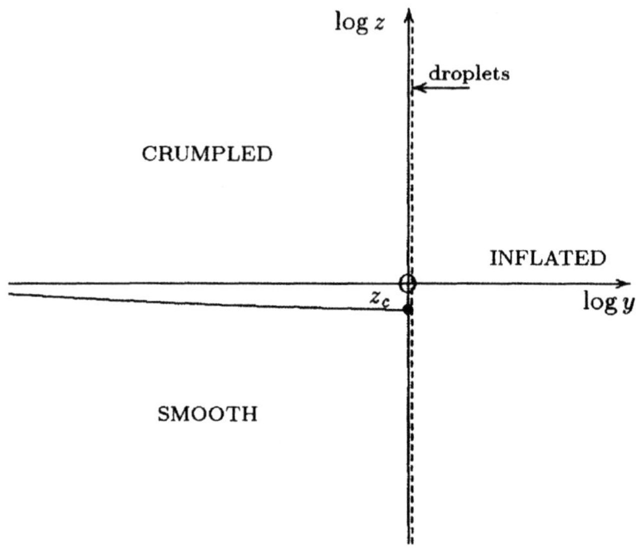

FIG. 3. *The phase diagram of c-surfaces. There are three phases coexisting at the triple point* $(0, z_c)$. *The axis* $\log z$ *is a line of first order transitions to inflated c-surfaces. A second line of crumpling transition separate the crumpled and smooth phases.*

are encountered. The third is at $y < 1$, were two phases of *deflated* smooth and *deflated* crumpled vesicles are encountered. These phases coexists at a triple point $(0, z_c)$.

5. Acknowledgements. The author expresses gratitude to the IMA for support during a three month visit (April to June 1996), and to NSERC Canada for an Operating Grant.

REFERENCES

[1] R. LIPOWSKI, *Nature* **349** (1991), 475.
[2] S.G. WHITTINGTON, *J. Math. Chem.* **14** (1993), 103.
[3] S. LEIBLER, in *Statistical Mechanics of Membranes and Surfaces*, eds. D. NELSON, T. PIRAN AND S. WEINBERG, *Jerusalem Winter School for Theoretical Physics* **5** (1987), 46 (World Scientific: Singapore, 1988).
[4] A.L. STELLA, E. ORLANDINI, I. BEICHL, F. SULLIVAN, M.C. TESI AND T.L. EINSTEIN, *Phys. Rev. Lett.* **64** (1992), 3650.
[5] E. ORLANDINI, A.L. STELLA, T.L. EINSTEIN, M.C. TESI, I. BEICHL AND F. SULLIVAN, *Phys. Rev. E* **53** (1996), 5800.
[6] E.J. JANSE VAN RENSBURG, *Preprint.*
[7] M.E. FISHER, A.J. GUTTMANN AND S.G. WHITTINGTON, *J. Phys. A: Math. Gen.* **24** (1991), 3095.
[8] E. ORLANDINI AND M.C. TESI, *Physica A* **185** (1992), 160.
[9] A. BAUMGÄRTNER, *Physica A* **192** (1993), 550.

[10] E. Orlandini, A.L. Stella, M.C. Tesi and F. Sullivan, *Phys. Rev. E* **48** (1993), R4203.
[11] E. Hille, *Functional Analysis and Semi-Groups (AMS Colloq. Publ.)* **31** (1948) (New York: AMS).
[12] J.B. Wilker and S.G. Whittington, *J. Phys. A: Math. Gen.* **12** (1979), L245.
[13] E.J. Janse van Rensburg and S.G. Whittington, *J. Phys. A: Math. Gen.* **22** (1989), 4939.
[14] M.E. Fisher, *Physics* **3** (1967), 255.

Current Volumes:

1 **Homogenization and Effective Moduli of Materials and Media** J. Ericksen, D. Kinderlehrer, R. Kohn, and J.-L. Lions (eds.)

2 **Oscillation Theory, Computation, and Methods of Compensated Compactness** C. Dafermos, J. Ericksen, D. Kinderlehrer, and M. Slemrod (eds.)

3 **Metastability and Incompletely Posed Problems** S. Antman, J. Ericksen, D. Kinderlehrer, and I. Muller (eds.)

4 **Dynamical Problems in Continuum Physics** J. Bona, C. Dafermos, J. Ericksen, and D. Kinderlehrer (eds.)

5 **Theory and Applications of Liquid Crystals** J. Ericksen and D. Kinderlehrer (eds.)

6 **Amorphous Polymers and Non-Newtonian Fluids** C. Dafermos, J. Ericksen, and D. Kinderlehrer (eds.)

7 **Random Media** G. Papanicolaou (ed.)

8 **Percolation Theory and Ergodic Theory of Infinite Particle Systems** H. Kesten (ed.)

9 **Hydrodynamic Behavior and Interacting Particle Systems** G. Papanicolaou (ed.)

10 **Stochastic Differential Systems, Stochastic Control Theory, and Applications** W. Fleming and P.-L. Lions (eds.)

11 **Numerical Simulation in Oil Recovery** M.F. Wheeler (ed.)

12 **Computational Fluid Dynamics and Reacting Gas Flows** B. Engquist, M. Luskin, and A. Majda (eds.)

13 **Numerical Algorithms for Parallel Computer Architectures** M.H. Schultz (ed.)

14 **Mathematical Aspects of Scientific Software** J.R. Rice (ed.)

15 **Mathematical Frontiers in Computational Chemical Physics** D. Truhlar (ed.)

16 **Mathematics in Industrial Problems** A. Friedman

17 **Applications of Combinatorics and Graph Theory to the Biological and Social Sciences** F. Roberts (ed.)

18 ***q*-Series and Partitions** D. Stanton (ed.)

19 **Invariant Theory and Tableaux** D. Stanton (ed.)

20 **Coding Theory and Design Theory Part I: Coding Theory** D. Ray-Chaudhuri (ed.)

21 **Coding Theory and Design Theory Part II: Design Theory** D. Ray-Chaudhuri (ed.)

22 **Signal Processing Part I: Signal Processing Theory** L. Auslander, F.A. Grünbaum, J.W. Helton, T. Kailath, P. Khargonekar, and S. Mitter (eds.)

23 **Signal Processing Part II: Control Theory and Applications of Signal Processing** L. Auslander, F.A. Grünbaum, J.W. Helton, T. Kailath, P. Khargonekar, and S. Mitter (eds.)
24 **Mathematics in Industrial Problems, Part 2** A. Friedman
25 **Solitons in Physics, Mathematics, and Nonlinear Optics** P.J. Olver and D.H. Sattinger (eds.)
26 **Two Phase Flows and Waves** D.D. Joseph and D.G. Schaeffer (eds.)
27 **Nonlinear Evolution Equations that Change Type** B.L. Keyfitz and M. Shearer (eds.)
28 **Computer Aided Proofs in Analysis** K. Meyer and D. Schmidt (eds.)
29 **Multidimensional Hyperbolic Problems and Computations** A. Majda and J. Glimm (eds.)
30 **Microlocal Analysis and Nonlinear Waves** M. Beals, R. Melrose, and J. Rauch (eds.)
31 **Mathematics in Industrial Problems, Part 3** A. Friedman
32 **Radar and Sonar, Part I** R. Blahut, W. Miller, Jr., and C. Wilcox
33 **Directions in Robust Statistics and Diagnostics: Part I** W.A. Stahel and S. Weisberg (eds.)
34 **Directions in Robust Statistics and Diagnostics: Part II** W.A. Stahel and S. Weisberg (eds.)
35 **Dynamical Issues in Combustion Theory** P. Fife, A. Liñán, and F.A. Williams (eds.)
36 **Computing and Graphics in Statistics** A. Buja and P. Tukey (eds.)
37 **Patterns and Dynamics in Reactive Media** H. Swinney, G. Aris, and D. Aronson (eds.)
38 **Mathematics in Industrial Problems, Part 4** A. Friedman
39 **Radar and Sonar, Part II** F.A. Grünbaum, M. Bernfeld, and R.E. Blahut (eds.)
40 **Nonlinear Phenomena in Atmospheric and Oceanic Sciences** G.F. Carnevale and R.T. Pierrehumbert (eds.)
41 **Chaotic Processes in the Geological Sciences** D.A. Yuen (ed.)
42 **Partial Differential Equations with Minimal Smoothness and Applications** B. Dahlberg, E. Fabes, R. Fefferman, D. Jerison, C. Kenig, and J. Pipher (eds.)
43 **On the Evolution of Phase Boundaries** M.E. Gurtin and G.B. McFadden
44 **Twist Mappings and Their Applications** R. McGehee and K.R. Meyer (eds.)
45 **New Directions in Time Series Analysis, Part I** D. Brillinger, P. Caines, J. Geweke, E. Parzen, M. Rosenblatt, and M.S. Taqqu (eds.)

46 **New Directions in Time Series Analysis, Part II**
D. Brillinger, P. Caines, J. Geweke, E. Parzen, M. Rosenblatt, and M.S. Taqqu (eds.)
47 **Degenerate Diffusions**
W.-M. Ni, L.A. Peletier, and J.-L. Vazquez (eds.)
48 **Linear Algebra, Markov Chains, and Queueing Models**
C.D. Meyer and R.J. Plemmons (eds.)
49 **Mathematics in Industrial Problems, Part 5** A. Friedman
50 **Combinatorial and Graph-Theoretic Problems in Linear Algebra**
R.A. Brualdi, S. Friedland, and V. Klee (eds.)
51 **Statistical Thermodynamics and Differential Geometry of Microstructured Materials**
H.T. Davis and J.C.C. Nitsche (eds.)
52 **Shock Induced Transitions and Phase Structures in General Media** J.E. Dunn, R. Fosdick, and M. Slemrod (eds.)
53 **Variational and Free Boundary Problems**
A. Friedman and J. Spruck (eds.)
54 **Microstructure and Phase Transitions**
D. Kinderlehrer, R. James, M. Luskin, and J.L. Ericksen (eds.)
55 **Turbulence in Fluid Flows: A Dynamical Systems Approach**
G.R. Sell, C. Foias, and R. Temam (eds.)
56 **Graph Theory and Sparse Matrix Computation**
A. George, J.R. Gilbert, and J.W.H. Liu (eds.)
57 **Mathematics in Industrial Problems, Part 6** A. Friedman
58 **Semiconductors, Part I**
W.M. Coughran, Jr., J. Cole, P. Lloyd, and J. White (eds.)
59 **Semiconductors, Part II**
W.M. Coughran, Jr., J. Cole, P. Lloyd, and J. White (eds.)
60 **Recent Advances in Iterative Methods**
G. Golub, A. Greenbaum, and M. Luskin (eds.)
61 **Free Boundaries in Viscous Flows**
R.A. Brown and S.H. Davis (eds.)
62 **Linear Algebra for Control Theory**
P. Van Dooren and B. Wyman (eds.)
63 **Hamiltonian Dynamical Systems: History, Theory, and Applications**
H.S. Dumas, K.R. Meyer, and D.S. Schmidt (eds.)
64 **Systems and Control Theory for Power Systems**
J.H. Chow, P.V. Kokotovic, R.J. Thomas (eds.)
65 **Mathematical Finance**
M.H.A. Davis, D. Duffie, W.H. Fleming, and S.E. Shreve (eds.)
66 **Robust Control Theory** B.A. Francis and P.P. Khargonekar (eds.)
67 **Mathematics in Industrial Problems, Part 7** A. Friedman
68 **Flow Control** M.D. Gunzburger (ed.)

69 **Linear Algebra for Signal Processing**
A. Bojanczyk and G. Cybenko (eds.)
70 **Control and Optimal Design of Distributed Parameter Systems**
J.E. Lagnese, D.L. Russell, and L.W. White (eds.)
71 **Stochastic Networks** F.P. Kelly and R.J. Williams (eds.)
72 **Discrete Probability and Algorithms**
D. Aldous, P. Diaconis, J. Spencer, and J.M. Steele (eds.)
73 **Discrete Event Systems, Manufacturing Systems, and Communication Networks**
P.R. Kumar and P.P. Varaiya (eds.)
74 **Adaptive Control, Filtering, and Signal Processing**
K.J. Åström, G.C. Goodwin, and P.R. Kumar (eds.)
75 **Modeling, Mesh Generation, and Adaptive Numerical Methods for Partial Differential Equations** I. Babuska, J.E. Flaherty, W.D. Henshaw, J.E. Hopcroft, J.E. Oliger, and T. Tezduyar (eds.)
76 **Random Discrete Structures** D. Aldous and R. Pemantle (eds.)
77 **Nonlinear Stochastic PDEs: Hydrodynamic Limit and Burgers' Turbulence** T. Funaki and W.A. Woyczynski (eds.)
78 **Nonsmooth Analysis and Geometric Methods in Deterministic Optimal Control** B.S. Mordukhovich and H.J. Sussmann (eds.)
79 **Environmental Studies: Mathematical, Computational, and Statistical Analysis** M.F. Wheeler (ed.)
80 **Image Models (and their Speech Model Cousins)**
S.E. Levinson and L. Shepp (eds.)
81 **Genetic Mapping and DNA Sequencing**
T. Speed and M.S. Waterman (eds.)
82 **Mathematical Approaches to Biomolecular Structure and Dynamics**
J.P. Mesirov, K. Schulten, and D. Sumners (eds.)
83 **Mathematics in Industrial Problems, Part 8** A. Friedman
84 **Classical and Modern Branching Processes**
K.B. Athreya and P. Jagers (eds.)
85 **Stochastic Models in Geosystems**
S.A. Molchanov and W.A. Woyczynski (eds.)
86 **Computational Wave Propagation**
B. Engquist and G.A. Kriegsmann (eds.)
87 **Progress in Population Genetics and Human Evolution**
P. Donnelly and S. Tavaré (eds.)
88 **Mathematics in Industrial Problems, Part 9** A. Friedman
89 **Multiparticle Quantum Scattering With Applications to Nuclear, Atomic and Molecular Physics** D.G. Truhlar and B. Simon (eds.)
90 **Inverse Problems in Wave Propagation** G. Chavent, G. Papanicolau, P. Sacks, and W.W. Symes (eds.)
91 **Singularities and Oscillations** J. Rauch and M. Taylor (eds.)

92 **Large-Scale Optimization with Applications, Part I: Optimization in Inverse Problems and Design**
L.T. Biegler, T.F. Coleman, A.R. Conn, and F. Santosa (eds.)

93 **Large-Scale Optimization with Applications, Part II: Optimal Design and Control**
L.T. Biegler, T.F. Coleman, A.R. Conn, and F. Santosa (eds.)

94 **Large-Scale Optimization with Applications, Part III: Molecular Structure and Optimization**
L.T. Biegler, T.F. Coleman, A.R. Conn, and F. Santosa (eds.)

95 **Quasiclassical Methods**
J. Rauch and B. Simon (eds.)

96 **Wave Propagation in Complex Media**
G. Papanicolaou (ed.)

97 **Random Sets: Theory and Applications**
J. Goutsias, R.P.S. Mahler, and H.T. Nguyen (eds.)

98 **Particulate Flows: Processing and Rheology**
D.A. Drew, D.D. Joseph, and S.L. Passman (eds.)

99 **Mathematics of Multiscale Materials** K.M. Golden, G.R. Grimmett, R.D. James, G.W. Milton, and P.N. Sen (eds.)

100 **Mathematics in Industrial Problems, Part 10** A. Friedman

101 **Nonlinear Optical Materials** J.V. Moloney (ed.)

102 **Numerical Methods for Polymeric Systems** S.G. Whittington (ed.)

103 **Topology and Geometry in Polymer Science** S.G. Whittington, D. Sumners, and T. Lodge (eds.)

104 **Essays on Mathematical Robotics** J. Baillieul, S.S. Sastry, and H.J. Sussman (eds.)

GPSR Compliance

The European Union's (EU) General Product Safety Regulation (GPSR) is a set of rules that requires consumer products to be safe and our obligations to ensure this.

If you have any concerns about our products, you can contact us on ProductSafety@springernature.com

In case Publisher is established outside the EU, the EU authorized representative is:

Springer Nature Customer Service Center GmbH
Europaplatz 3
69115 Heidelberg, Germany

Batch number: 10704041

Printed by Printforce, the Netherlands